The Companion Guide to Trackards for North American Mammals

To Olaus Murie,
the inspiration,
who led the right life.

The Companion Guide
to
Trackards
for
North American Mammals

by

David Brown

The McDonald & Woodward Publishing Company
Newark, Ohio

The McDonald & Woodward Publishing Company
Newark, Ohio
www.mwpubco.com

The Companion Guide to
Trackards for North American Mammals

Printed in the United States of America by
McNaughton & Gunn, Inc., Saline, Michigan, on paper that
meets the minimum requirements of permanence
for printed library materials.

Second printing January 2021

10 9 8 7 6 5 4 3 2
30 29 28 27 26 25 24 23 22 21

Table of Contents

Acknowledgements

I wish to thank Kevin and Rita Harding for their kindness, friendship and support for many years. Kevin has been my companion on many tracking excursions in both New England and Arizona. His separate point of view and insights have been invaluable in developing this project. Kevin's wife, Rita Harding, has also been invaluable in keeping my body and soul connected through many lean years with her generosity and mastery of the kitchen. Gus, their beloved cat, should also be thanked for his patience in tolerating the probing of his paw pads for insight into the structure of the feline foot. I have known no other cat in my life that would stand for it!

My friend, Joe Choiniere, has also been a companion on many tracking trips. He is an excellent writer and the best general naturalist I know. His service in reviewing the final manuscript before submission was invaluable. His wife, Donna Nothe-Choiniere, an excellent naturalist in her own right, also reviewed the manuscript and offered suggestions that are gratefully received.

Over many years of conducting tracking and other wildlife programs I have had the pleasure of meeting hundreds of participants. Their perspectives and questions have led to insights into better ways of explaining what are often complex processes of identification and interpretation of tracks, trails and other sign. I am indebted to them for these questions and perceptual insights.

Tracker and photographer Paul Rezendes provided early guidance way back in the 1980s and early 90s. He was the first person I had ever met who had the competence to confirm or deny authoritatively my intuitions about tracks and sign. I might never have gotten started on this long journey of discovery without his help.

Olaus Murie was a biologist for the U.S. Biological Survey, which later became the US Fish and Wildlife Service. In his inspiring life he accomplished many firsts. He was the first biologist I know of who looked at the natural history of the West with a careful eye while it still was wild, that is, before it was converted to fenced cow pastures, oil fields and industrial timberlands. His book, *A Field Guide to Animal Tracks*, was the first authoritative rendition of tracks and sign in print. In addition his tracking adventures have inspired many, myself included, as much for his privileged access to a disappearing landscape as for his careful renditions of the tracks and sign of what he saw. From noting the tooth marks on wildflowers outside a cabin in the Rockies to dogsledding down the Yukon River, he had an opportunity to live the right life and took full advantage of that opportunity.

The bibliography at the back of this work lists books that I have consulted at one point or another over twenty-five years and have found useful in advancing my own knowledge. I am greatly indebted to these authors who have gone before and shown the way.

Finally, I would like to thank Dr. Jerry McDonald, of McDonald and Woodward Publishing, for taking a chance on *Trackards for North American Mammals* and *The Companion Guide to Trackards for North American Mammals*. I should also thank him and his staff for their hard work in converting to commercial publication what was for many years a self-produced guide. Thanks as well to Trish Newcomb, their marketing director, for her efforts in making both publications available to appropriate points of sale. I hope the future for this project will reward their confidence and efforts.

The Companion Guide
to
Trackards
for
North American Mammals

Introduction

At the beginning of presentations I sometimes tell the story of my first tracking experience as a boy. In the marshy bed of Penny Brook in Saugus, Massachusetts, I discovered in a dusting of snow over ice the prints of a small animal that had passed that way in the night. I studied the trail carefully, trying to imbed its pattern and the appearance of its individual tracks in my memory. Later, I took down the small collection of tracking books at the local library, intent on identifying what I had seen. I was excited by the prospect of discovering the secret life that the trail represented, concealed from human eyes in the dark. I hadn't expected much difficulty; there were, I thought, a limited number of possibilities at a site close to a major eastern city where few wild animals were likely to live. However, though I scoured the books, each of which generally displayed a single pen and ink drawing per species, I could find nothing that seemed to resemble what I had witnessed. Excitement was reduced to frustration, and with the dimming of memory by time, I have no real idea today what species made the tracks I saw on that hike through the woods many years ago. As I tell this story, I often see smiles of recognition on the faces of my listeners. Other people who are interested in nature, it seems, have had a similar experience.

What was the problem? Why didn't those tracking guides lead me to a successful identification? A foot is a foot, after all, and the impression of the foot of any animal of one species should be similar to that of any other member of that species, shouldn't it? Well, perhaps it should, but it rarely does, and on that hangs a tale.

How many times have we heard "It's easy and fun!" in advertising? Ease and pleasure are so thoroughly linked in the modern mind that the publishers of many guides are understandably wary of presenting complexity, even if complexity is called for by the subject matter. A tracking guide, reduced by a concise format to minimal information, simply won't be useful in the field. An animal's tracks and trails are subject to all sorts of distortions so that a single representation of a track or trail is likely to be a misrepresentation. Most of the time little that is found in the field will resemble the single illustration in the book. As a result, the book guides its user nowhere, and the window into the lives of wild animals that flee from us and hide in the night remains shut.

What you hold in your hands is not one of the oversimplified "quick and easy" guides. It attempts to present tracks and tracking in something closer to their true complexity. But it also tries to organize that complexity in such a way that success will result more often than frustration. No single book could ever contain in the finite space within its two covers all possibilities since no two tracks are ever exactly alike. But most animals leave tracks that are variations on a theme, and identifying that theme through multiple illustrations of common appearances is the method of this guide. Tracking is rarely easy. But mastering the complexity of what is at least a minor art is one of the satisfactions that I find in the activity. If it were easy, I would have lost interest long ago.

There are three parts to a track: your mind, the track itself, and the being that made it. Let's start backwards. The animal whose foot impression you find in the snow is capable of moving in lots of different ways. It can use what we of Western education reduce to a half dozen gaits, each of which has a somewhat different weight distribution and impact. With each gait a track will take on a somewhat different appearance. The track of a walking fox in firm mud will look delicate. However, if the same fox is bounding, the high arc of its motion results in greater impact: the toes spread slightly, the pads flatten and expand, the secondary pad may be de-emphasized in the impression and the deeper track will take on a more robust appearance. The two tracks will bear a resemblance, to be sure, but they will be variations of a fundamental shape. Members of another species, raccoons, are one of the great foolers. They can spread their toes or close them, extend or contract them, and make their tracks look like those of several other animals. No single representation of their tracks could possibly suffice. Snowshoe hares are the other of the great foolers: they can spread their toes from a minimum of 1¾ inches all the way out to 4 inches!

An animal track itself is imbedded in a surface, and the possible variations in a surface are infinite. There are innumerable varieties of snow, for instance, each of which will present a somewhat different appearance when a foot is impressed in it. Add different consistencies of earth, mud and sand, and the possibilities are limitless. Furthermore, tracks are laid down in patterns corresponding to a half-dozen gaits that can blend into one another, confounding the neat Western mind with so many variations that the usefulness of the categories themselves are called into question.

Finally, the mind of the observer needs to be taken into account. A track is capable of infinite detail depending on how closely we look and how much of that detail we choose to observe. The mind is set the task of reducing that detail to something manageable, to decide what is significant detail and what isn't. This is the editing function of the mind. All our minds do it, but each does it differently, and that's the problem. We bring our entire genetic and cultural heritage to bear on anything that we look at. And we all have different heritages. In a sense, what we already know dictates what we will know. And we all already know different things. As a result no two observers will see exactly the same track with the mind's eye. Try tracing the photograph of a track, as I have done many times for this guide. What appeared to be a perfectly defined track at first glance begins to slip away. We discover that where the mind saw edges and completed ellipses, the pen is staggered with indecision. The edges fade, the ellipses vanish. Again and again one must back up and look with the mind to decide where in a gradient of gray the outline should go. The result is invariably a compromise of sorts. Where will the general mind see an edge? Will the general human mind see a whole where there isn't one? The difficulty is, of course, that there is really no general human mind. The mind of a primitive, trying to depict a creature on a cave wall, often saw that creature with distortions that we find incredible. Hopefully, the reader and myself are culturally near enough to one another that the way I see a track will be close enough to the way he or she will see it that an accurate identification can be made. Field-testing of the *Trackards for North American Mammals* and *The Companion Guide to Trackards for North American Mammals* over many years suggests that this is usually the case.

Well, now you've heard the worst. If you persist, then rewards await. I often liken the trail of a wild animal to a diary of its daily, or more likely nightly, activities. Observed directly, a wild animal will usually be aware of the observer and alter its behavior accordingly, normally by running away at top speed. The knowledge you gain will be the appearance of its rear end disappearing. But a trail of prints, set down when the animal was unaware of observation, contains candid information about its behavior that may be read after the fact by the tracker, who then is in a position to reconstruct the event in his imagination with a level of accuracy corresponding to his skill and experience. In this way a connection is made from the human mind, through the track and trail, to the body of the wild animal and ultimately to its wild and primitive consciousness. Perhaps in this devious way we can arrive at an awareness of the wildness buried deep in our own sophisticated, civilized, rationalizing mind, as well, and make the connection to our own elusive and increasingly remote natural context, a place that we seem to be seeking when we feel the need to go to the woods.

From the reconstructed event comes both practical and ethereal knowledge. Why did the mink follow the shoreline? Why did the fisher change gaits in the snow? What diverted the attention of the fox as it trotted across a frozen Walden Pond, leaving a hiccup in its otherwise elegant trail for Thoreau to ponder from his rustic cabin? Speculations on the answers to these things bring us closer to the inner being of the animals we follow. They also provide the ecological connection: what was there about this place and time that attracted the animal here? The more we practice, the more accurate our surmising becomes and the more the hidden world opens to us.

But the text of that diary is written in a foreign language and will tell little to its reader until he has learned some of its vocabulary and syntax. Much of that grammar is for another book, however. The animal must first be identified. The *Trackards* and *Companion Guide* are intended to facilitate that identification and help the reader on his way to an adventure that will enhance his natural awareness and enrich his life.

The *Trackards* and *Companion Guide* attempt to improve upon the accuracy, usefulness and usability of other track identification guides in a number of ways. Little did I realize, back in that library when I was a boy, how fanciful were the drawings of many of the tracks in those books as their authors filled in gaps in their experience from their imagination. The illustrations on the *Trackards* and in *The Companion Guide*, on the other hand, were traced and shaded directly either from photographs or from plaster casts of actual tracks and sign made by the weight-bearing feet of live, free-ranging wild animals. Photographs representative of each species were selected from a collection of nearly eighteen hundred slides and digital images acquired during twenty-five years of fieldwork and from a large collection of plaster casts collected over the same period. The individual images were chosen from the collections according to their ability to communicate the fundamental forms typical of each species.

The current cards and book present the fourth revision of this work. Where the earlier images were mostly traced from projected slides, often in soft pencil to allow for corrections, and then scanned, a number of images in this revised edition were made from digital photographs. These were printed and then traced with a fine pen on a light-box before being scanned and fitted to scale on the page or card. As a result of the different processes of their production some slight inconsistency in the texture of the outlines and shading on the images may be detected: those done by pencil and photocopier tend to be a little softer than the newer ones done with the digital-pen combination. Nonetheless, this inconsistency of style has no effect on the functionality of the results. Finally, illustrations made from track casts were traced on a transparent plastic medium, then converted to pen and ink before scanning and scaling. Whatever the process, the resulting illustrations, like the primitive's hand traced on the cave wall, are as free from the distortions of memory, imagination and individual perspective as I can make them. They are, I am confident, an improvement in accuracy and authenticity over most other track identification guides available.

One might ask why the photographs themselves are not presented on the *Trackards*. While photographs of tracks, if they are well done, are certainly accurate, all too often they are gray shadow on white snow or brown on brown earth. In the field the parallax of the eyes and their changeable perspective can detect the depth of a low-contrast track and bring it into relief in the mind. The camera's image, however, is two-dimensional and has a fixed perspective, flattening the track against its background. In effect, although we view an actual track in three dimensions, we are restricted to two in a photograph. The illustrations on the *Trackards*, on the other hand, transfer vague photo impressions to high-contrast black and white and render them with depth shading to create the illusion of the missing third dimension.

Illustrations that were considered most likely to be useful for side-by-side comparison with sign found in the field were selected and are presented, actual size, on the *Trackards* rather than being relegated to *The Companion Guide*. In some cases idealized versions are presented, as well, for clarity. While some useful notations are provided on the cards, the user may refer to *The Companion Guide* for additional explanation of the Trackard illustrations as well as for additional examples of tracks and sign. Since no two tracks are exactly the same, and in some cases can vary widely, the multiple examples are intended at least to span the possibilities. Studying them at length may imbed a fundamental shape in the mind that represents a consistent theme in the many variations ("gestalt"), a sense that then can be carried to observations in the field.

Finally, in the trails section at the bottom of each card are presented various track patterns that each animal typically uses. These patterns are interpreted in the text for gaits and behavior, as well.

Both the *Trackards* and *Companion Guide* are sized to fit into a large jacket pocket. Each card is made out of a synthetic material that is impervious to moisture. Unlike the illustrations in a bound book, they can be placed without harm on mud or snow for direct comparison of dimensions and appearances. The cards are ringed loosely and the tracks of similar size are grouped together in the deck. By fanning a few cards and placing them on the ground, full-sized images can be isolated against the actual tracks.

The information on the *Trackards* and in *The Companion Guide* comes from personal experience as recorded on film, in plaster and in the pages of my field notebook. Every effort has been made to avoid mistakes and to indicate theory, when it is presented, as opposed to fact. However, animal tracking is an emerging art rather than an exact science. If this book contains errors, at least I can say they are honest ones. Designed for field use, the set is intended to supplement (and upon rare occasions correct) the fine and honest work of the pioneer, Olaus Murie, whose pen and ink drawings have inspired so many others since, sometimes in less than honest ways, and of my contemporary, Paul Rezendes, whose photographs are just too beautiful to put down on the ground.

Ambiguities provide little more than frustration; one learns little from the trail of an animal one cannot first identify, and so all too often the interested person is brought to a standstill at the point of beginning. While having someone at hand with enough experience to confirm or deny our intuitions is ideal, few of us can have such a resource constantly at our elbow. I hope and trust the *Trackards* and their *Companion Guide* will serve as a helpful substitute.

How to Use the Trackards

The *Trackards for North American Mammals* are intended for rugged use in the field. Each card is printed on synthetic, waterproofed stock so that it can be placed directly on the ground for comparison of the illustrated tracks and sign, actual size, with those found in mud, snow and sand. The cards are easily removed from the ring, if necessary, but they are ordered in such a way that this should seldom be necessary. The order of the cards follows the size of the animals' prints. In cases where the front and hind prints have different sizes, the order follows the larger of the two.

The assumption in ordering the *Trackards* is that the user will first come upon an unfamiliar track that he wishes to identify, as opposed to examining a Trackard and then going in search of the animal represented. Proceeding from this assumption, the *Trackards* are arranged in descending order of approximate track size: Trackard 1 displays the tracks and scat of a bear, which has the largest prints in the set, while Trackard 25 shows the tracks of voles and shrews, the smallest included. When one finds a track that needs to be identified, simply estimate whether it is larger or smaller than that of a medium-size animal such as a woodchuck. If it is larger, it will be found on the top side of the deck, somewhere between Trackard 1 – Bear and Trackard 13 – Woodchuck. Hold the deck so that the Bear card is on top and then fan the cards slightly so that the index tabs on their lower corner show all the animals of woodchuck track size or larger. If the mammal has a track smaller than that of a woodchuck, turn the set over so that Trackard 26 – Miscellaneous is on top and fan the cards to show the index tabs of smaller animals from Trackard 14 – Red Fox down in size to Trackard 25 – Vole and Shrew.

This arrangement allows the tracker to find all the prints that surround the size of the track in question among consecutive cards on one side of the deck. Furthermore, to prevent confusion, the back of each card shows the tracks of an animal that are nowhere near the size of the tracks on the front. This eliminates the need to turn cards over, and it allows the user, if he or she chooses, to remove the ring and place on the ground face-up for comparison all the cards that are possibilities.

Along one edge of the card is an inch scale that may be used as a ruler to measure the discovered sign. Measurements typical of adults of the species are indicated on this ruler with a central line from the caption box indicating the peak of the distribution and a fan to suggest its breadth. The span is deliberately vague since there can be variation in size for different ages, sexes, regions of the country and even different individuals. Measurements will also vary according to the surface in which tracks were impressed. For instance, a track in soft, dry snow or breakable crust will often measure wider than it would in mud or damp snow. As you may already know or will soon learn, little in the art of tracking lends itself to perfect consistency. In the beginning the novice will rely a lot on measurements because he needs something definite and quantitative upon which to hang his identification. As his experience accumulates, the tracker will gain more confidence in his intuitions, and reliance on measuring will fade. This is not to say that measuring is useless to the experienced tracker, only that it is just one factor of varying importance in a preponderance of evidence upon which an accurate identification is made.

At the bottom of most cards is a section that shows track patterns frequently found in the trail of the species. Since different mammals favor particular gaits over others in various circumstances, the ones most typical of the animal are presented. Additional patterns may be included in *The Companion Guide*. In illustrations where the prints are presented as circles or ovals, the darkened ones represent front prints and the light ones show the hind. In "direct registrations," where the hind print coincides with the front, the circle is divided diagonally, half white and half black. The same warnings about reliance on measurements mentioned above apply as well to step lengths, pattern widths and straddles presented in this section. Some are more consistent and useful than others; all represent an average around which there is always some distribution. An alertness to the conditions in which the trail was laid down will help you adjust upward or downward from the presented values. For instance, most young animals are growing up to independence in late summer and fall. During this period undersized tracks and step length can be expected, especially since at this season young animals greatly outnumber adults.

On the major portion of each Trackard is presented the information that the user needs to make a comparative identification in the field. Thus, on all the cards at least one print or set of prints along with, in most cases, a representative scat are presented full size so that a direct comparison can be made as to size and shape by placing the card on the ground. In some cases, a full-size idealized rendition of prints is also included for clarity. Other illustrations that present tracks in different mediums, depths or toe arrangements, each of which presents a somewhat different appearance from that of the idealized or common prints, are presented in *The Companion Guide* along with examples of other sign. The scat of many species, for instance, often varies widely with season or diet and so selected alternate appearances are illustrated in *The Companion Guide*. On some cards browse sign may be presented as well. This is especially the case with cuttings that are frequently seen and easily confused, such as between rabbit and deer or between beaver and porcupine. To facilitate reference, the chapters in *The Companion Guide* are ordered in the same size-sequence as the Trackards.

It should be obvious that illustrations on the Trackards will look too large when the cards are held at reading distance, waist level or above. Viewing them on the ground, as they are intended to be used, will dispel the illusion. Wiping them occasionally with a damp cloth or sponge will eliminate any sticky residue that may accumulate from such field use and make them easier to fan and separate.

Trackards for North American Mammals and *The Companion Guide to Trackards for North American Mammals* have been field tested for over a decade by hundreds of users and have been revised several times. The resulting sets, if used as they were designed, should allow positive identifications most of the time. Although no book or system can expect to match the infinite variety of Nature, a solid understanding of how this identification system works along with careful observation and a little imagination should result in success.

Glossary

An effort has been made to reduce the amount of specialized jargon in *The Companion Guide* and to use common words as they are generally understood rather than assigning them esoteric meanings. Nonetheless in any field there will be some terms for things not anticipated in everyday speech. Below is a glossary of helpful tracking terms as they are used in this book. These terms are divided into three sections: General Terms, Pads and Gaits. Two figures provide visual assistance for better understanding details of pads (Figure G 1) and gaits (Figure G 2).

General Terms

Aligned: a gait in which the hind foot on each side moves directly over the print of the front foot on that side. In an aligned gait the animal's spine is arranged parallel to the direction of travel. See "Displaced" below.

Alternating: a zigzag trail of tracks representing a walking or aligned trotting gait.

Alternating slant: tracks arranged on a diagonal relative to the direction of travel but with the diagonals reversed from pattern to successive pattern. See "Repeating slant" below.

Arc of stride: the arc described by following a point on an animal's torso, viewed from the side, through one cycle of gait motion (a full "stride"). Walking has a flat arc of stride while bounding has a high arc and may be referred to as having a lot of "verticality."

Coincident: in this guide, any gait in which the front and hind feet of an animal rise and fall together through the cycle of a stride. Trotting, for instance, is a coincident gait. See "Rolling" below.

Concave: curved inward; think of a "cave." Opposite of "Convex" below.

Convex: curved outward like the surface of a ball or the dome of a hilltop. Opposite of "Concave" above.

Digitigrade: moving either on the toes alone or on the toes and "secondary pad(s)." See "Plantigrade" below.

Displaced: in this guide, a gait in which the hind foot moves past rather than over the print of the front foot that is on the same side. To do this the animal displaces its hind end to one side, holding its spine at a slant relative to the direction of travel. See "Aligned" above.

Even: tracks or pads arranged side-by-side and perpendicular to the direction of travel.

Flat: having a low arc of stride, that is, a gait having insignificant vertical motion.

Fractionate: to take apart a scat in order to investigate its contents.

Gestalt: here, the fundamental form of a track or trail that survives variations and may be used to identify the animal.

Lateral: toward the side, rather than the centerline, of an animal's body when viewed from above. See "Medial" below.

Medial: toward the centerline, rather than the side, of an animal's body when viewed from above. See "Lateral" above.

Pattern width: the distance across a track pattern from the lateral edge of the leftmost print to the lateral edge of the rightmost print, measured perpendicular to the direction of travel. It is synonymous with "trail width." However, note the difference between this term and "Straddle" below.

Phalangials: toe bones, or the linear toes (digits) they make up, the impressions of which sometimes show in a print between the toe pads and the secondary pad(s) of a foot.

Plantigrade: moving on the entire sole of the foot, including the "tertiary pad" or area. See "Digitigrade" above.

Primary pads: toe pads. Sometimes called "digital pads." See Figure G 1 and discussion under "Pads" below.

Print: in this book, the impression made by a single foot. See "Track." In some other tracking guides these terms are synonymous.

Registration: the process of creating an impression with the foot. "Single registration" (SR) occurs when the print of each foot appears without overlap or contact with any other print. "Direct registration" (DR) occurs when a hind print coincides perfectly with a front. "Indirect registration" (IR) means that the hind foot landed slightly off-center from the front print, leaving a single distorted track. "Double registration" (DblR) occurs when the hind print overlaps the front but so widely off-center that the presence of two prints in the track is clear. Note that other tracking guides use the term "double registration" to refer to two single registrations that are simply close together in a pattern.

Repeating slant: tracks arranged on a diagonal relative to the direction of travel and repeating the diagonal from pattern to successive pattern rather than reversing it, as is the case in "Alternating slant" above.

Retarded: located toward the rear of a print or pattern; the opposite of "advanced." In this guide "retarded" often refers to a print feature that is located toward the rear relative to the same feature on the opposite side, giving the print an asymmetrical appearance.

Rolling: any gait with a rocking horse motion in which the front end of the animal is up when the hind is down and vice versa. Loping, galloping and bounding are rolling gaits. See "Coincident" above.

Rotary: a four-print (4X) "rotary" pattern has an order of placement: right-left-left-right or left-right-right-left. Wider patterns have the shape of the letter C or a comma, thus the term "C-pattern." See "Transverse" below.

Scat: the droppings of a wild animal. Scat is measured at the point of maximum diameter. Averaged measurements are shown on the edge of the Trackard.

Secondary pad(s): the pad or pads that register in a print immediately to the rear of the toe prints. See the illustration and discussion later in this chapter. See Figure G 1 and discussion under "Pads" below.

Sign: any evidence for the former presence of an animal. Although tracks may be considered sign, the term is usually reserved for such things as scat, digs, rubs, chews, bones, hair, scent marks and so forth.

Signature: a mark or outline appearing in part of a print that, while not always present, definitely identifies a species when it does appear.

Step length: in this guide, a measurement from any point in the print of one front foot to the same point in the print of the other front foot (or hind foot to hind foot) in the course of a single step. Step length is measured parallel to the direction of travel. In slow gaits, like walking and trotting, step lengths are more or less constant; however, in rolling gaits, like loping and galloping, step lengths (unlike "strides") usually alternate short and long. See "Stride" below.

Straddle: in this guide, the distance across the two front or hind prints measured from the lateral side of each and perpendicular to the direction of travel. In "displaced" gaits, the straddles of both front and hind prints are narrower than the "pattern width" or "trail width." Note that other tracking guides may use all of these terms synonymously.

Stride: the distance between two consecutive prints of the same foot. A stride amounts to two consecutive "step lengths" and is more or less constant along a trail in which the animal did not vary its gait. Note that other guides often use "stride" and "step length" synonymously.

Suspension: any period in the stride cycle where all four of a mammal's feet are off the ground. In "extended suspension" the animal's body is stretched out in mid-leap; in "gathered suspension" the animal's hind feet have moved forward past its front feet so that front and hind limbs form an X under the body when viewed from the side.

Tertiary pad(s): generally any pads in a print that register to the rear of the secondary pad(s). See Figure G 1 and discussion under "Pads" below.

Track: in this book, a single entity that may be the print of only one foot or the partially or perfectly overlain impressions of more than one foot.

Track width: the width of a track measured perpendicular to the direction of travel. Usually this is from the medial edge of the inner toe pad to the lateral edge of the outer toe pad.

Trail width: normally synonymous with "pattern width" above, but in some short-legged animals in snow, such as porcupines and otters, body or fur impressions may make the trail width wider.

Transverse: a four-print (4X) "transverse" pattern has the order of placement: right-left-right-left or left-right-left-right. Wide and long patterns have a shape more or less like the letter T, thus "T-pattern." Short patterns are often called "1-2-1." See "Rotary" above.

Vestigial: a toe pad or other foot part that is gradually disappearing in the course of evolution. Many wild animals have a "vestigial" fifth toe, located on the medial edge of the foot, that registers only slightly or not at all in a print. The human "little toe" is vestigial but is located on the lateral side of the foot, the opposite of those mammals that are included in *Trackards*.

Withdrawn: held above the general plane of a print. A withdrawn pad is de-emphasized in a track, registering lightly or not at all. The medial fifth toe of a black bear is often withdrawn from prints. Compare "Retarded" above.

X (as in **2X, 3X** and **4X):** indicates the number of tracks evident in a pattern. Note that these may range from direct to single registrations; thus, a 2X pattern shows only two "tracks" but these may be the impressions (prints) of 2 or 4 feet, or more in the case of two animals following in each other's footprints. In a 3X pattern, one hind foot lands in the impression of one of the front feet, producing a single track to go along with the prints of the other two feet.

Pads

Several competing systems have been used by different authors to designate the pads on various animals' feet. For one reason or another none of these systems seems adequate, being either too complex, vague, illogical or inconsistent from animal to animal. At the risk of making an already confusing situation even more so, a simple nomenclature, which will be used in this guide, is advanced below. Whatever it lacks in anatomical precision, it has proven adequately descriptive for fieldwork and is much simpler and more consistent than other methods now available on the market.

Primary pads: toe pads, the first grouping of pads on an animal's front or hind foot. Also called "digital pads."

Secondary pad(s): the second pad or cluster of pads, located immediately to the rear of the toe pads. When there is more than one secondary pad, they are usually contiguous, or at least visually in the same group, and separate from the primary pads ahead and tertiary pads to the rear. In animals with fused footpads such as bears, skunks and humans, where there is no clear division between secondary and tertiary pads, the "secondary area" of the sole of the foot is referred to, meaning the breadth of the foot in the area of the human ball. This nomenclature works best for soft padded animals. Hoofed animals show dewclaws in the secondary area, but dewclaws are really the first and fourth toes of the animal's foot, which have moved back up the limb in the course of evolution, and thus are technically "primary." In the case of ungulates (hoofed animals) then, my usual nomenclature is suspended in favor of the traditional "cloves" and "dewclaws."

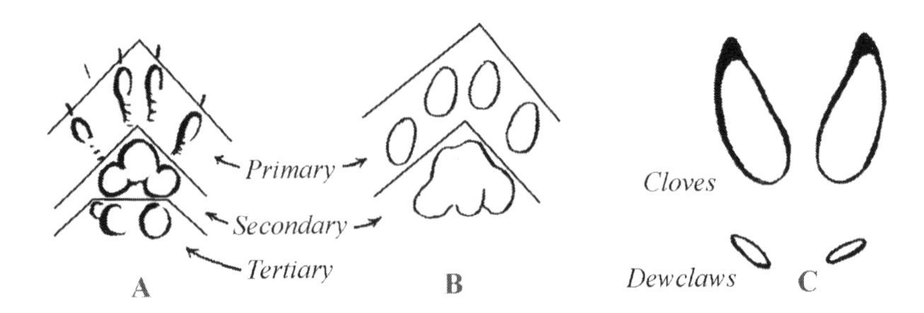

Figure G 1. Pads, cloves and dewclaws.

A. A track with primary, secondary and tertiary pads. Note the phalangial marks between the primary pads and secondary pads.

B. A track with primary and secondary pads. Phalangial marks are not present.

C. A track of a deer showing the cloves of the hoof and the vestigial dewclaws.

Tertiary pad(s): any third pad or grouping of pads that shows on the heel area of a print. Some animals have them and some do not. Humans and other plantigrade walkers do; digitigrade walkers like dogs, cats and deer don't, at least not positioned so that they appear in prints of a moving animal. Some animals may register a tertiary area in the heel without pad separation from the secondary area, such as in the hind print of a bear. Tracks of other species, such as opossums, made when the animals are at rest may show a heel area but no recognizable pad, while the hind prints of squirrels may have small scattered heel pads that only rarely register.

The remaining parts of a mammal's foot structure that sometimes appear in a print are the phalangials which connect the toe pads to the secondary pads. In mammals with naked feet, that is, without inter-digital fur, the impression of these bones is sometimes diagnostic in identification. See, for example, Fisher versus Otter.

It should be noted that animals which normally register four toe pads in a print, like dogs and cats, have a vestigial medial fifth toe that has migrated up the foot into the tertiary area. However, this toe seldom shows in a track and so the problem it poses in this simplified terminology will be ignored. It should also be noted that the foot of such animals actually extends up to the angular joint that we intuitively think of as the animal's elbow. But the heel of this foot has no fleshy pad, since it is seldom in contact with the ground. The bottom line is that dog, cat and deer/moose prints show no tertiary pad and, unless the animal is sitting down, no tertiary area either.

Gaits

Walk: in a quadruped, a gait in which one foot moves forward at a time leaving the other three in contact with the ground (Figure G 2). Through most of the stride cycle tripod stability is maintained. There is no "suspension" in a walking gait.

Trot: a slow run in which two feet, one front and one hind, are in contact with the ground while the other two are in synchronized motion. The pairings may be on the same side of the body as in a "trotter" at a race track or, more usually with wild animals, diagonally opposite. A trot always involves at least a brief moment of suspension. Trotting is "coincident," that is, the shoulders and haunches rise and fall together.

Lope: an easy run in which two feet (two front or two hind) are in contact with ground while the other two are moving forward. This is a "rolling" gait: when the shoulders are down, the haunches are up and when the haunches are down the shoulders are up. There may be brief moments of both gathered and extended suspension. Most propulsion comes from the hind legs, the front being used mostly as props.

Gallop: an energetic, rolling gait, similar to a lope in its mechanics. However, in a gallop the animal propels itself with its front legs as well as its hind. Although there are always moments of both extended and gathered suspension, the body is held low and with a low arc of stride, maximizing the period in which the feet are in contact with the ground. Wild horses and pronghorns are the only wild North American mammals that can sustain a gallop for any distance; for others it is a brief attack or escape gait.

Bound: also an energetic rolling gait with mechanics similar to a lope. However, bounding has a high arc of stride. This verticality makes it inefficient as a way to get around on flat, even terrain. Most animals use it to clear soft snow, tall grass or debris on the forest floor.

Hop-bound: a gait used by many rodents bounding through soft snow. The front feet land and pre-pack the snow for the hind feet that follow in and at least partially cover the front. There is usually a slight pause while the animal extricates front from under hind and

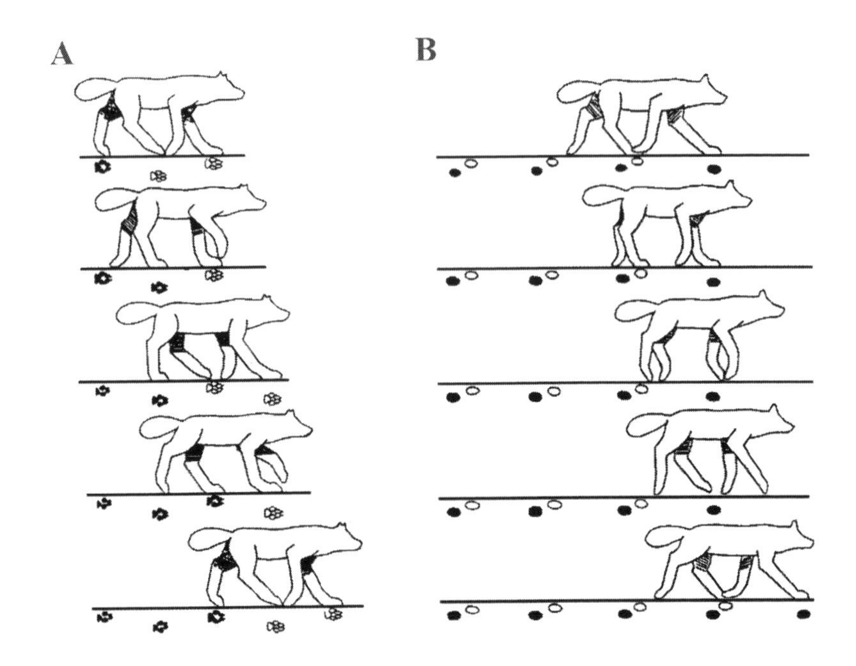
A
B

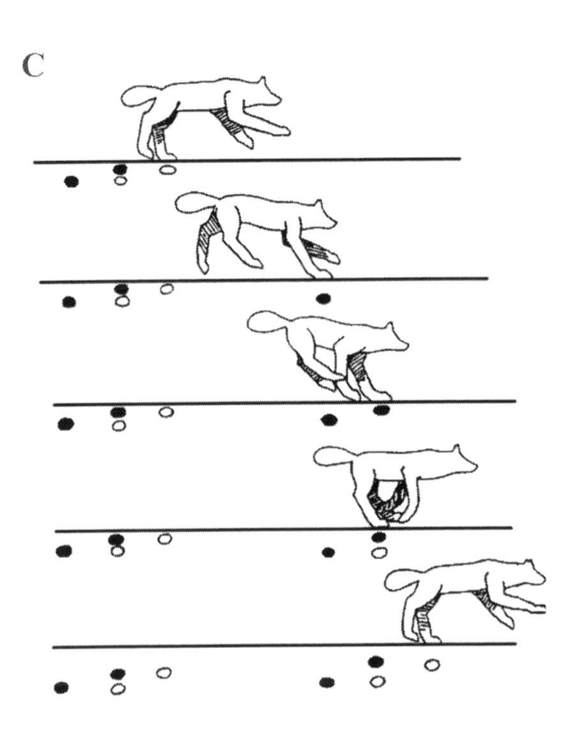
C

launches itself with its powerful hind legs into another short, high arc.

Lope-bound: an easy run with a higher arc of stride than a pure lope and a tighter grouping of prints. Lope-bounding is used, often by long legged animals, for snow or low debris, alternating with other gaits such as pure loping or bounding as changes in the surface are encountered.

Run: a general term for any locomotion faster than a walk.

Scurry: a fast walk or trot by a small mammal.

Figure G.2. Gaits and track patterns.

A. Walking, showing the "alternating DR" pattern, a zigzag trail of alternating direct registrations. In similar patterns the hind feet may understep, registering behind the front; overstep, registering ahead of the front; or indirect register laterally.

B. Trotting (displaced), showing the "repeating slant 2X SR" pattern in which two repeated prints arranged on a slant that repeats from pattern to pattern rather than alternating. Note that in this pattern, common among canids, all the front prints are retarded and lined up on one side of the trail while all the hind prints are advanced and lined up on the other.

C. Lope-bounding, showing the "transverse 1-2-1" pattern that begins with a single front print followed by a pairing of front and hind prints arranged even or on a slant and finished with a single hind print. The order of placement here is "transverse," that is, in the order of right-left-right-left.

Trackard 1 – Black Bear

There are three species of bears in North America: the polar bear, the brown (or "grizzly") bear and the black bear. In the contiguous United States the grizzly only occurs in the northern Rocky Mountains, while the black bear inhabits most of the country. Black bears of the mid-Atlantic states can grow quite large in places where there is a mature hardwood forest with a lot of mast. In the West, on the other hand, this species is usually fairly small. The various bear species leave similar footprints, so that distinguishing among them is largely a matter of size, range and habitat. The black bear of New England is presented as intermediate in size for its species.

With an arched "Roman nose" and lacking a shoulder hump, a black bear is fairly easy to distinguish by sight, at least, from its normally larger and fiercer cousin, the grizzly. The tracks of the two species, however, are harder to tell apart. The fur color of a black bear is, as its name implies, normally black, with brown around the muzzle, although other color morphs are occasionally seen. With the regrowth to maturity of the eastern forest, black bears, which are the arboreal counterpart of the terrestrial grizzly bear of the American West, are becoming increasingly common. Unfortunately this brings them more and more into contact and conflict with humans. Intelligent, resourceful and stubborn, they have learned what a red and white plastic box is for and can be alarmingly insistent on its possession. Keep campground coolers out of reach – and a bear's reach can be long! Recent reports show that some large bears have learned that they can gain access to food in cars by stamping with their front feet against a window until it breaks. If a bear insists on taking your food, it is best not to argue the issue.

Tracks

Figure 1.1 A shows a right hind print, full size, from a cast made on Popple Mountain in New Hampshire in April snow. This print is

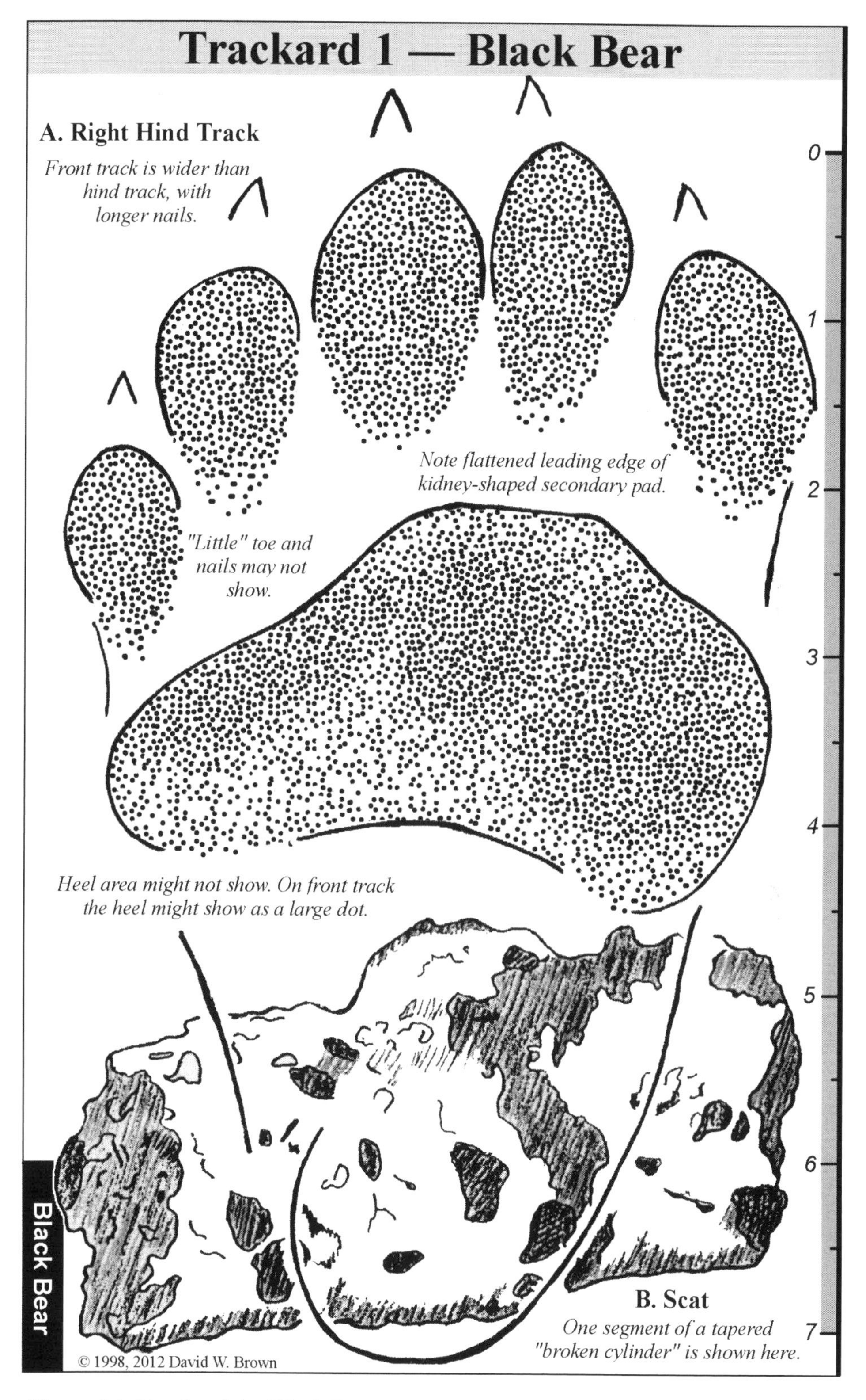

Figure 1.1. Trackard 1 – Black Bear.

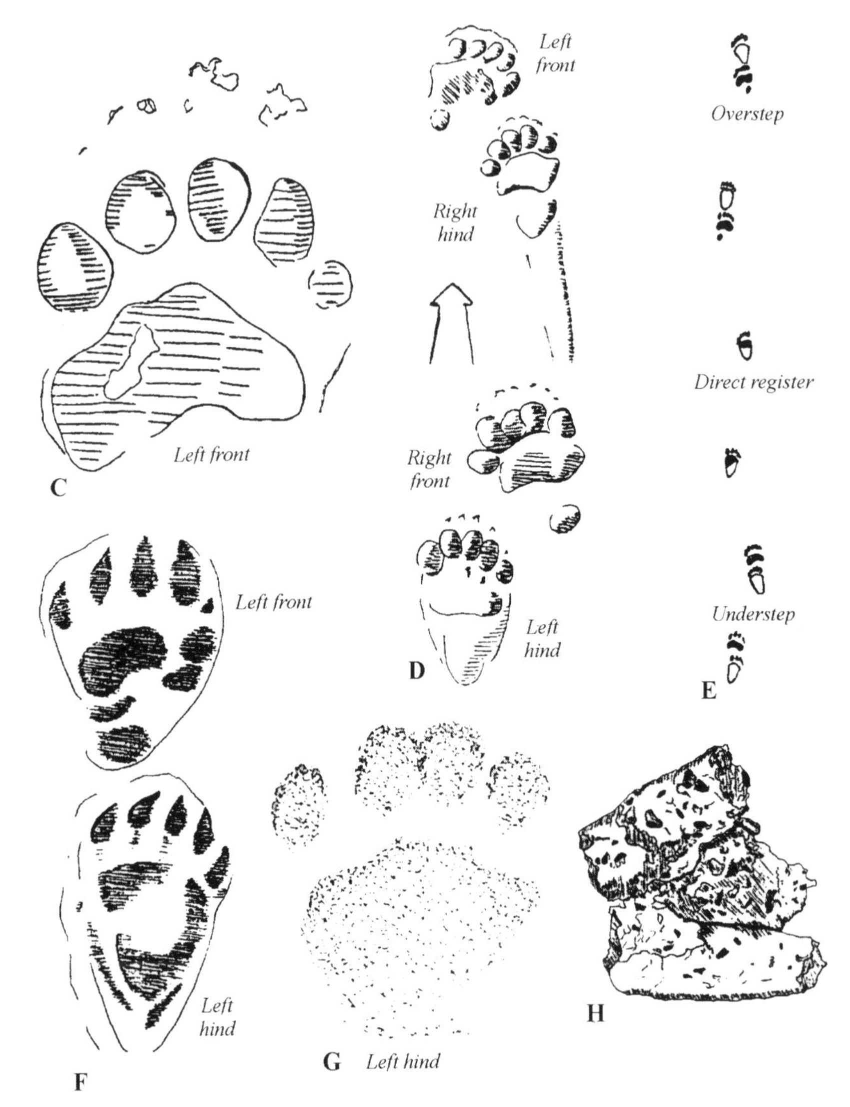

Figure 1.2. Black bear tracks, trails and scat.
C. Front print in snow, depth 2 inches. Petersham, Massachusetts.
D. Walk in snow with large overstep, step length 25–30 inches. Jackson, New Hampshire.
E. Walk/trot patterns, speeding up. Step length 16–28 inches. Jackson, New Hampshire.
F. Front (above) and hind prints in snow, depth 3 inches. Jackson, New Hampshire.
G. "Cat" print in streamside sand. Passaconaway, New Hampshire.
H. Apple scat, maximum diameter 2¼ inches. Jackson, New Hampshire.

about average for New England bear tracks, which normally measure between 4 and 5 inches wide. Black bears have five toes on both front and hind. The "little toe," which is on the medial side of the foot, is normally de-emphasized, registering lightly and in shallow tracks sometimes not at all. Its location is also more or less retarded toward the rear in many prints. A slight space often appears between the lateral "big toe," which is actually about the same size as the middle toes, and the other four toes. The secondary pads are fused into one smooth area that may be slightly emphasized on the lateral side. In contrast to humans, then, the "ball" of a bear's foot is on the other side, the lateral side, of the foot. Although bears are very flat-footed, a vague "arch" may appear in some tracks.

Although generally plantigrade walkers, moving on both ball and heel of the foot, bears are not always and necessarily so. Like humans, they can move "on their toes" in which case the heel will not register at all and the resulting print will be composed only of toes and the secondary area. This is the case in Figure 1.2 G, a hind print recorded on a sandbar in the Swift River of northern New Hampshire at a depth of less than an inch. In this instance the medial 5th toe also failed to register and the nails, which are under muscular control, were also withdrawn above the general plane of the pads. The result is a four-toed impression, without nails or heel, which looks rather cat-like and quite different from the print on Trackard 1.

Figure 1.2 F on the same page shows two snow prints of a male just out of hibernation in the White Mountains of New Hampshire. They were about 4 inches deep in soft spring snow and illustrate the deformability of a soft-padded animal's foot.

Front print: In most cases the front print appears shorter and wider than the hind. The front heel (tertiary area) normally registers as just a dot (Figure 1.2 D) or not at all. The secondary area is kidney-shaped. Nails are longer on the front than on the hind and often register more prominently.

Hind print: The hind print usually appears longer and narrower than the front and is proportioned more like that of a flat-footed human. The secondary and tertiary pads are fused into one large smooth pad that usually registers completely although in faster gaits where the bear is more up on its toes, the tertiary may be de-emphasized and disappear entirely from the track as in Figure 1.2 G.

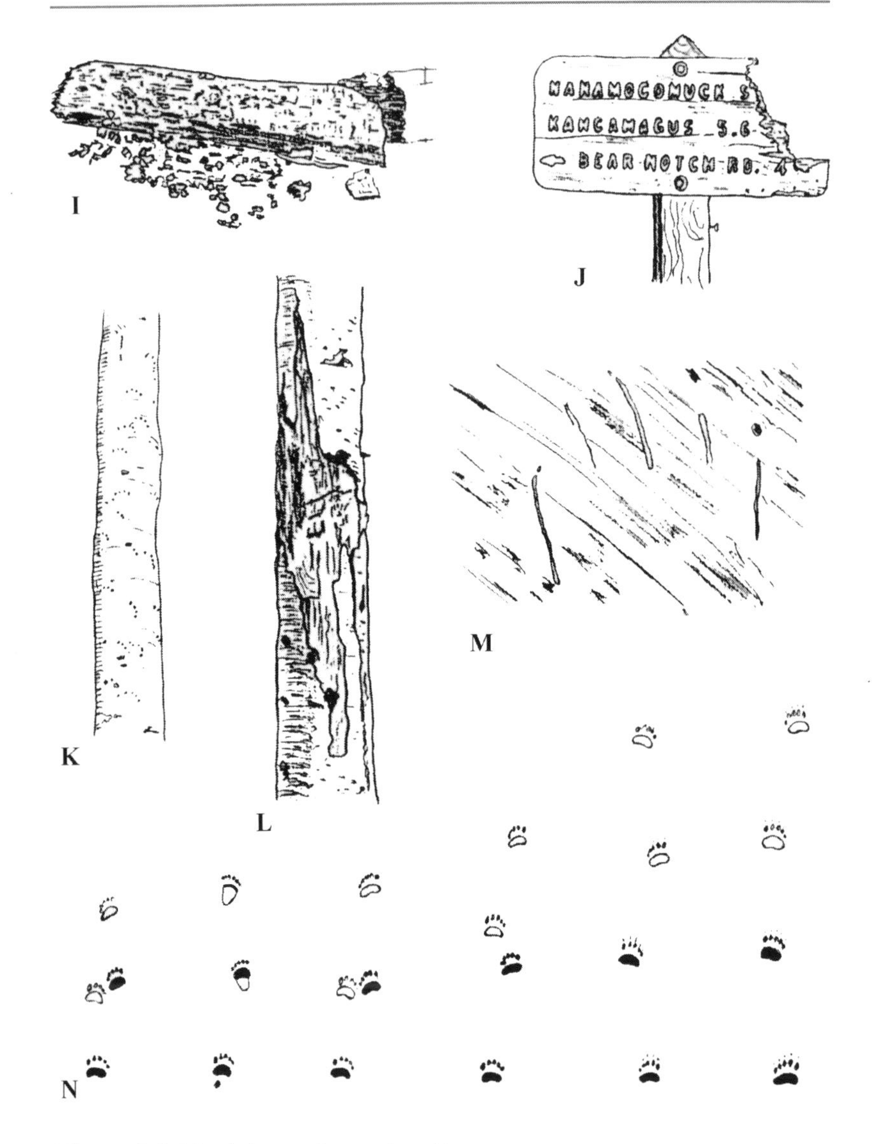

Figure 1.3. Black bear sign and patterns.

I. Log torn apart for grub and beetles. Passaconaway, New Hampshire.

J. Bear "sign." Passaconaway, New Hampshire.

K. Climbing marks on beech. Jackson, New Hampshire.

L. Balsam fir marked by bear, 6-inch diameter. Passaconaway, New Hampshire.

M. Claw marks on log across dry streambed. Passaconaway, New Hampshire.

N. Lope and gallop patterns speeding up from left to right.

Trails

Black bears walk a lot. The pattern may show direct registrations, especially in deep tracks in soft snow where such a habit improves efficiency. Often, however, a walking bear will show an understep in which the hind print registers to the rear of the front, or an overstep where it registers ahead of the front print. A line of these patterns appears in Figure 1.2 E where the bear transitioned from a slow understep to direct registering to a faster overstep. Direct registrations as well as oversteps may represent walking but may also be aligned trotting (see Glossary). When the gait is a walk, an adult black bear in the Northeast normally shows a step length between 18 and 23 inches with a lot of toe-in. As the gait speeds up to a trot, the straddle typically narrows, the feet straighten out and the step length can increase to as much as 33 inches. The overstep both in a walk and trot can increase to such an extent that the hind print actually registers just behind the front print on the other side as in Figure 1.2 D, recorded on Maple Mountain in April snow with an average step length of 25–30 inches. When the animal is not trotting I call this gait a "wiggle walk." It is also commonly used by both raccoons and bobcats. All three are tree-climbers with very mobile pelvises that allow for this sort of gait distortion.

Bears can also lope and, for short distances, even gallop. Figure 1.3 N shows an assortment of such patterns recorded in the Passaconaway intervale in New Hampshire and the Quabbin area of central Massachusetts with speed increasing from left to right. Bears may lope after jumping an obstacle, for instance, or to avoid observation when they are crossing a road, or when climbing or descending a hill. Picture in your mind a "lumbering bear" with rolls of autumn fat rippling under its fur and you are picturing this loping gait. Galloping may be used as an escape gait by a bear that knows it has been observed or has sensed danger. Pattern length will be longer, to about 5 feet, than in a lope; nail marks will show prominently, as may loosened debris behind both front and hind prints.

Trails that are repeatedly used by bears moving through the forest will be padded smooth, not lumpy like those of hoofed animals. Sometimes bears on frequently-used trails are reported to step in the same prints over and over again until depressions form, spaced evenly along the trail, although I have only rarely found this in the Northeast. In boreal regions, check the height of any breakage of the dead branches on the lower part of conifers in order to help distinguish bear trails from those of taller animals like moose. Where bears are hunted, their sign is usually found well away and upslope from human trails in the forest. However, in thick growth like the stunted spruce slopes on the upper parts of eastern mountains, bears sometimes opt for the easier going of hiking trails. The passage of a bear, even over autumn leaves, is miraculously soundless. Walking humans kick up fallen leaves, walking bears pad them down.

Track and Trail Comparisons

Mountain lion: When a bear leaves an impression as in Figure 1.2 G, it is sometimes mistaken for a track of a mountain lion. With only four evident toes and a common flattening of the forward edge of the secondary pad the impression is superficially cat-like, and this common profile accounts for many false reports of mountain lions in the North Country. Note, however, that the arc of the bear's toes is much shallower than that of any cat, appearing in a row at the head of the print rather than surrounding the forward lobe of the secondary pad (Figure 1.2 G). Note also the absence of any scalloping on the rear edge of the bear's secondary pad, a feature that often appears clearly in the print of a cat (see Mountain Lion, page 237).

Grizzly bear: In some parts of the country with rich habitat, black bears, which are usually much less aggressive than grizzlies, can attain the size of their normally larger and fiercer cousins. However such ranges rarely overlap with grizzly habitat so that there should be a substantial size difference between adult grizzly and adult black bear prints in the same area. However, the prints of a young grizzly can be the same size as an adult black, and if this grizzly is still attached to its mother, trouble may follow. Normally the arc of the toes of a grizzly is shallower than that of a black bear and the nails longer, especially in the front print, a sign of the grizzly's dependence on digging as opposed to the shorter climbing claws of a black bear. However, be careful! A foot is a living and flexible thing and a bear's nails are under muscular control; either can be distorted by a number of factors that may overwhelm the expected differences between these two species.

Sign

Logs softened by rot are often torn apart by bears looking for beetles, grubs and other insects as in Figure 1.3 I from the Swift River in New Hampshire. Such logs are visited repeatedly. Sometimes standing deadwood is knocked down to expose the rotting roots for the same purpose as well as to extricate voles that may be tunneling there.

So-called "bear nests" are loose clumps of branches broken from the canopy of beech stands and collected in one of the upper forks of a tree. These are associated with feeding, not sleeping, however, and might more properly be called "bear ricks." As the animal breaks off branches to get to the nuts at their tip, it sticks each limb into the collection of branches under its feet to steady itself and control the branch for feeding.

Bears leave a variety of signs on trees in their range. Usually these involve raking the tree downwards with front nails, rubbing the back on the roughened wood and biting the inner wood. Figure 1.3 L from Passaconaway shows such work. In this region bears seem to prefer balsam firs about 6 inches in diameter for this treatment. Red pines of any size are also favorites. On remote roads without a lot of traffic bears also work on telephone poles in this way. In Figure 1.3 L note the typical bite showing diagonal marks of the canine teeth. A pocket magnifier may reveal hair stuck to the sap around the wound as well. The height of claw marks on such trees is not necessarily a good indicator of bear size as young bears may jump up before digging in their nails. The height and breadth of the diagonal tooth marks are better indicators.

In thick vegetation, bears may use a dry streambed as a highway. Look for scat deposits and check tree trunks that have fallen across the bed. If the bear climbed over such a trunk, claw marks may appear on its upper surface as in Figure 1.3 M from the Swift River. If the bear squeezed underneath, it may also have left hair stuck to any roughening of the undersurface.

Bears often "vandalize" trail signboards, clawing and biting them as in Figure 1.3 J also from Passaconaway. Such treatment may be a reaction to something new in the bear's environment, the unusual taste of the paint or simply gum exercise. Toothmarks indicate that the bears often like to sink their teeth into trail signs a couple of inches thick and rock back and forth, loosening the post. Shims jammed into the ground at the base of trail signs by forestry workers are a frequent sight in bear country.

In late summer and fall respectively, bears climb fruit and nut trees, leaving claw marks which are particularly noticeable on the smooth bark of beeches as in Figure 1.3 K from Popple Mountain in

New Hampshire. It should be noted that the spread of these clawmarks is not relatable to the size of the animal unless the gouges are very fresh: as beech trees age after indentation, the marks widen and spread.

Sign Comparisons

Moose: Moose frequently gnaw the bark of sweet-sapped trees like maples and mountain ash. The long marks of their lower incisors are often mistaken for bearclaw marks. But the tails of curled bark at the end of the rake marks will be mostly at the bottom when a bear does it, at the top when the mark is a moose gnaw. Also, moose break saplings to browse the upper foliage and buds, not to get at berries or fruit as in bear work. Look for sign of what has been eaten.

Porcupine: Porcupines gnaw signboards for the taste of the paint, which they neatly and systemically remove from the underwood. Bear work tends to be more damaging to the underlying wood. Porcupines, like moose and bears, often break off the tops of small trees. But they do so by climbing so high that they overtop and ride the treetop to the ground. They are after buds, leaves or bark, not fruit. Look for bear clawmarks on the bark. Paired incisor marks close together anywhere above 3 feet on the tree mean porcupine; similar marks below 3 feet, consider beaver. Additional sign of porcupine includes its distinctive scat near the base of a gnawed tree or quills loosened and lost upon impact.

Scat

Classed as carnivores, black bears will indeed take prey animals up to the size of a fawn. However, they have adapted to a mostly vegetarian diet. Like most animals that feed on seasonally available fruits, nuts, farm corn and so forth, their scat can vary enormously. Spring droppings contain early greenery, such as skunk cabbage, and look like bundles of wound fiber. Also in the spring, bears will tear up large patches of fallen leaves looking for beechnuts that fell after the first snow of autumn and have remained suspended in the snowpack all winter. In this case the scat will show the indigestible husks of these nuts. Scat resulting from ants, a summer staple, will show shiny black exoskeletons on the surface as well, perhaps, as bits of rotted wood or other debris swallowed along with the ants. Fruit scat will have the color and odor of the source. Figure 1.1 B shows a single segment of a typical late fall scat, full size, and Figure 1.2 H shows the whole deposit. Because the bear had been in an orchard eating apples, the resulting scat was yellow and rust-colored, with pieces of apple skin on the surface and a cidery odor.

Whatever the source, bear scat is large. The illustrated scat is typical, a broken cylinder with a maximum diameter from about 1½ to 2½ inches. Where the bear was diarrheic from fruit, the scat will be a formless "bear pie," once again copious, containing the seeds of the fruit. Cherry scat will show a reddish pile of large pits, blueberry scat will be blue, blackberry black and so forth.

Bear urine can be copious as well, sometimes deposited absent-mindedly while the animal is walking along, leaving a lengthy stain. Its odor may vary with diet but is generally mild. I recall a large deposit of early winter urine at Petersham, Massachusetts, smelling something like coyote urine but without the musk.

Scat Comparisons

Eastern coyote: When Eastern coyotes are feeding on fall apples, their scat, which commonly measures over an inch in diameter, can be hard to tell from that of a small black bear. However, coyote scat is usually either teardrop-shaped or of uniform diameter for most of its length while many bear deposits show at least one piece which tapers smoothly for its entire length as in the bottom segment in Figure 1.2 H. Furthermore bears will usually come and go from an orchard through nearby woods where their scats are likely to be deposited at random. Coyotes are bolder, more often approaching and departing over roads or trails which they may deliberately mark with their scats. However, in thick or boggy country bears may use hiking trails or roads a great deal and their scat as well as tracks may be found along them. As is often the case in tracking, identifications are best made on the sum of the evidence, not on individual details. The presence of scratch marks on the ground or lightly musked urine nearby will, of course, point to the coyote.

Raccoon: The runny deposits left behind by raccoons, which also feed on ripened fruit, can be quite copious. Wide claw marks on fruit trees, broken branches and wide trails through berry patches are all circumstantial evidence in favor of bear work, however.

Habitat

Unlike their fierce cousins of the West, black bears are adapted to trees. Tracks may be found in spring and fall snow in hillside beech stands or extensive groves of other nut-bearing trees. In non-snow seasons, look for tracks in streambank sand and mud; since they are large and their dark fur is sun-absorptive, they drink a lot. Sign may be found wherever the omnivorous bears feed, especially orchards and berry patches in season, under nut trees, around campgrounds and along hiking trails. Ironically, mid-summer is the hardest time for bears as far as food gathering is concerned. Fruits and nuts have yet to ripen; the spring vegetation has "hardened up" and is indigestible. Bears use this season for mating because there is little else to do. Unfortunately this starvation time for bears is also the peak of camping activity for humans, and the lure of easy pickings at a campground or campsite may be too much for the normally reclusive bears to resist.

"Natural" sign at this season will include log-rolling in older clear-cuts in search of ants and grubs. Look carefully at the surface of a log that has been rolled out of place for the telltale four or five claw marks where the animal raked its claws over the surface until they caught on some irregularity that gave it purchase to pull the log over. Compressions of the wood along the leading edge of broken pieces may also show where the claws gained that purchase.

In winter, northern bears hibernate in a hollow in the ground or in a dead tree. More southerly bears may go into torpor with no more cover than some blowdown or a brush pile. If there is a plentiful winter berry crop such as mountain ash, bears may stay out of hibernation until very late into the season. Even winter campers need to be careful with their food!

An unfortunate consequence of our desire to feed birds in winter is that we supply an attractive food source to bears, which are good students of our habits and will return time and again to places where we have provided food in the past, even to the point of delaying hibernation to take advantage of them. Horse barns are also an attraction, their visits terrifying the horses stabled within. However, it is not the horses the bear is interested in, but rather what the horses eat. Once discovered, unprotected grain bins become a memorable item in a bear's brain. Lastly, bears have a well-earned reputation for a sweet tooth. Bee hives provide satisfaction for this craving, with bears becoming a serious nuisance to apiarists and a very difficult one to dissuade from nocturnal visits.

Trackard 2 – Beaver

Trapped almost to extinction during the 18th and 19th centuries for its beautiful and valuable pelt, the beaver has made a remarkable comeback in the East, reclaiming much of its ancient range and providing wetland habitat for a wide variety of other animals. It radically modifies its environment by building dams that impound water. The created pond provides safety from which the animal can forage to cut down nearby trees for food and building materials. By cutting trees out of the canopy it allows sunlight to reach the surface. Since sunlight and water are the principal ingredients of life on Earth, the created pond becomes a magnet for wild animals and plants of all description.

The beaver is a large semi-aquatic mammal; weights of as much as 60 pounds have been reported! The most distinctive feature of its anatomy is its broad naked tail, which serves to store fat, to regulate body temperature and to warn other beavers on the pond when it is slapped on the water to signal danger.

Tracks

Tracks are most often found in the mud around the beaver's lodge, on the muddy rim of its dam or in the trench that it wears down between the pond and cutting areas. Classic beaver tracks with an entire outline, such as those in Figure 2.1 A and B, recorded in snow at Passaconaway, New Hampshire, are seldom found. The broad tail usually erases part or all of the hind print, and both tail and huge hind foot normally obscure the small front print. Dragging cut branches down its trails also contributes to the obscurity of beavers' tracks. The result is often a vague impression in which neither toes, webbing nor heel area appears. Certain features often survive the erasure, however, and can prove diagnostic. One signature is the tendency of beavers to leave deep nail holes that often look like human fingernails. These can be seen on a few of the toe prints in Figure 2.2 E and G. The beaver's toes are so long that these must be looked for well ahead of the palm impression from which they often appear disassociated. Another signature is the impression of knuckle pads under the joints of the upper palm (the secondary pad area) of the hind foot, which often survive tail and branch dragging. In fluffy or melted-out snow sometimes the only distinctive characteristic discernible may be the concave curve that often outlines the sides of the hind print, giving the track the shape of an old-fashioned flared vase.

When front prints can be found, they often look like a child's hand with a finger missing as in Figure 2.1 B. Although beavers have a fifth toe, it seldom can be discerned on the rare occasions when the front print can be found at all. Figure 2.2 E shows a variety of front print profiles recorded in thin snow over ice on the Conway River in New Hampshire.

Trails

Beavers mostly walk. The trail appears as a series of alternating impressions in which the hind print understeps, oversteps or covers the front print. In early winter snow before the pond freezes over and while the beaver is putting the finishing touches on the cache of branches beside its lodge, trails often can be found leading from the water's edge to a cutting area and back. Figure 2.2 G shows the impressions of a front and hind in thin snow over bare ground in November at Petersham, Massachusetts, while Figure 2.1 D shows a walking trail in snow in late December by the Swift River in Passaconaway, New Hampshire. Even in deep winter beavers sometimes come on land in order to cut saplings to bolster a deficient winter cache or to wear down their teeth on a handy oak tree. In the trail shown on Trackard 2 (Figure 2.1 D), the front prints are covered by the hind, and tail drag is not evident; a beaver's tail radiates heat and so raising it off ice or snow reduces conductive heat loss and protects it from abrasion. Another place to look for the trail of a beaver in winter is on a road above a culvert connecting two flowages. Perhaps nervous about being trapped in the tube by a predator or being ambushed upon emerging, beavers seem to prefer going over rather through these constrictions.

Trail Comparisons

Otter: Viewed from a distance, from across a stream for example, a beaver trail on an embankment can look like an otter slide,

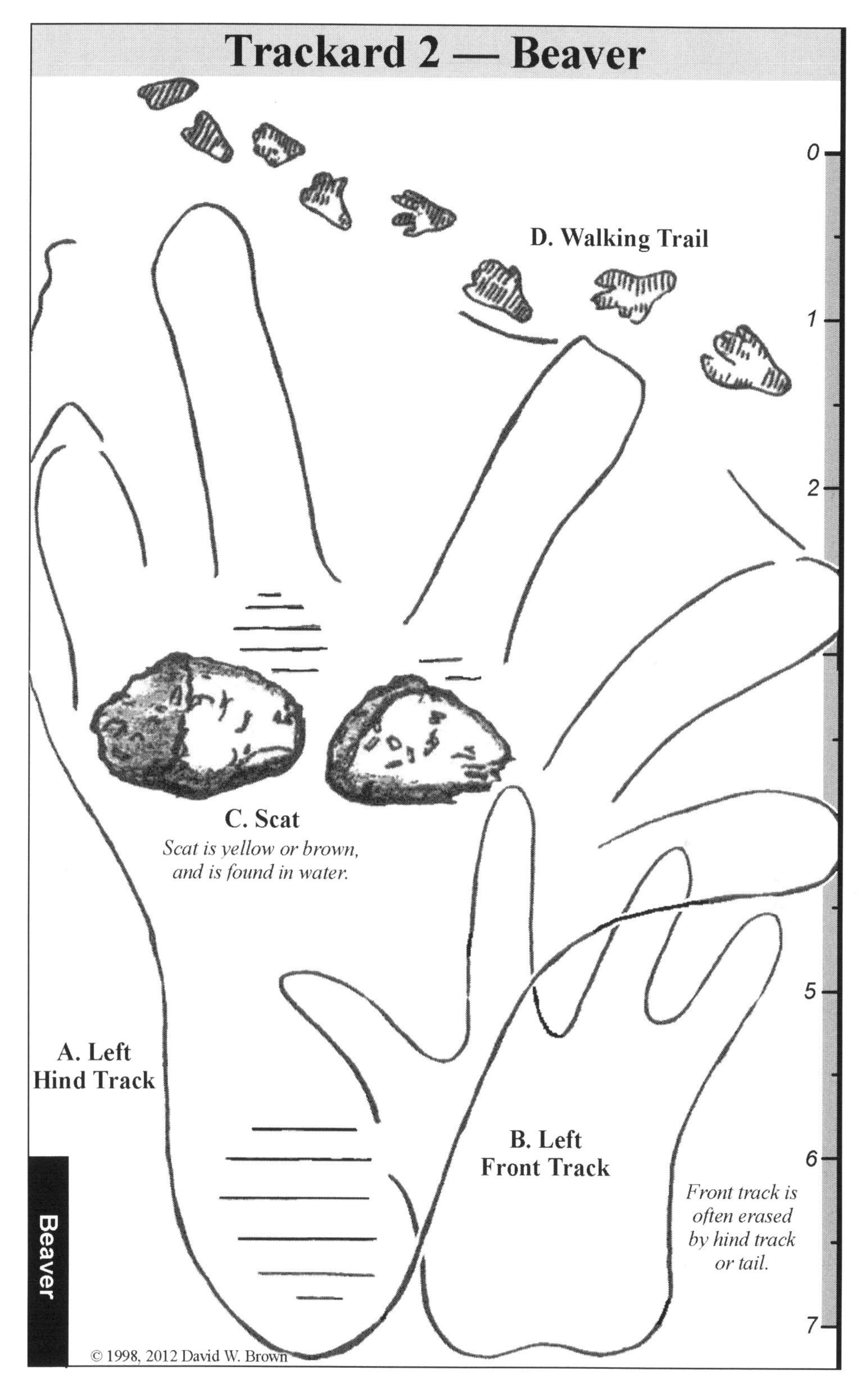

Figure 2.1. Trackard 2 – Beaver.

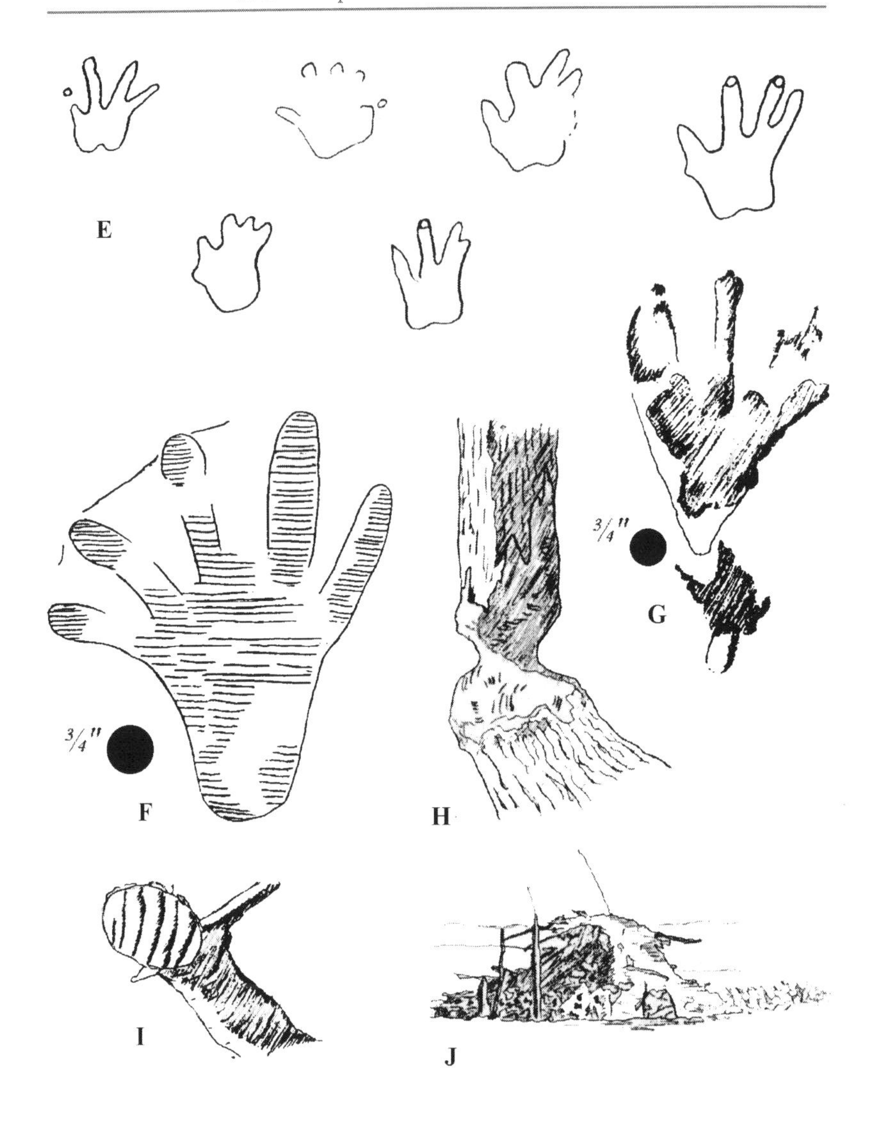

Figure 2.2. Beaver tracks and sign.

E. Front prints in thin snow over ice. Conway, New Hampshire.

F. Hind print in snow, depth 3 inches. Passaconaway, New Hampshire.

G. Front and hind prints in thin snow over bare ground. Petersham, Massachusetts.

H. Swamp maple cutting. Conway, New Hampshire.

I. Birch branch cutting. Petersham, Massachusetts.

J. Free-standing lodge. Petersham, Massachusetts.

and indeed sometimes the activities of both species can overlap one another. However, the trough of a beaver will often zizzag across the slope as the animal switchbacks its considerable bulk to the top. Where it has dragged branches straight down the slope to the water, look for snow disturbed at the edge of the trough by twigs or, in summer, disturbed leaf litter or grass combed downslope. Otters slide down the fall line usually over a narrow, well-defined trough into the water. Beaver trails generally look like work while otters' look like play.

Sign

While tracks of beavers are not commonly found, other sign of their presence is conspicuous. Beavers cut down trees near their ponds in order to feed on the cambium layer under the bark as well as to use the wood for construction. Aspens are usually the first to go, followed by sweet-barked trees such as black birch and often yellow birch as well. The red maple in Figure 2.2 H from the Conway River floodplain in New Hampshire is one of the less-favored species and may have been gnawed as potential building material as much as for food.

In winter beavers may periodically risk a trip ashore to gnaw hard-wooded trees like oaks or perhaps the maple in Figure 2.2 H, apparently as a way of wearing down their incisors, which grow continuously. Such trees can be differentiated from other cutting because the beaver gnaws the tree only part way through, clearly without intending to cut it down. Sometimes gnawed areas of different ages will be apparent by the different color of the wood as the beaver works on the tree over a number of winters.

A common sign of beaver occupancy is the scent mound. This can be as small as a wad of leaves and mud pushed up onto the bank of the pond or nearly as large as a muskrat dome. On top is deposited castoreum, a glandular substance with which the beaver proclaims its tenancy. Castoreum has a distinct and pleasant odor that permeates the area and will announce the beavers' presence to you as well as to others of their species.

Beavers make channels through the lily pads in their ponds, through early skim ice and through the emergent grasses and sedges on their margins. These can be distinguished from muskrat channels by their width of about 2 feet compared to the 6-inch-wide troughs of the smaller rodent.

Sign Comparisons

Porcupine: Beaver chewing can usually be told from that of porcupine quite easily. Porcupines do not gnaw the heartwood of a tree since they are uninterested in cutting it down to feed on the upper branches. They, unlike beavers, can climb up the standing tree to feed. Any extensive debarking that removes only the cambium layer and is much above ground- or snow-level can usually be attributed to cousin porcupine. However, beavers do sometimes gnaw the cambium of certain trees like hemlock and beech without gnawing into the heartwood. The proximity of more recognizable beaver activity in the vicinity as well as the absence of porcupine scat should distinguish the two.

Branches that a porcupine cuts while perched in the canopy can usually be told from beaver work as well. The diagonal cut of a porcupine shows a lot of splitting while beaver cuts are usually finely chiseled like the birch branch from Petersham, Massachusetts, shown in Figure 2.2 I. Also, porcupines rarely debark these branches, while beavers strip them of their bark and cambium extensively.

Lodge and Bank Burrows

The typical lodge of a beaver family is pictured in Figure 2.2 J from a pond in Petersham. It may be distinguished from the houses of muskrats by its size and by its construction with debarked sticks as opposed to the softer cattails and reeds that muskrats use. Lodges are renovated mostly in the late summer and fall in preparation for winter. Once the pond is frozen, coyotes and other predators will be able to venture over the ice to the lodge, so the beavers plaster it with fresh mud, which freezes solid into a nearly impregnable fortress. For food over the winter the beavers swim out of the lodge through an underwater entrance and cut off a meal from their winter cache of prepositioned branches stuck into the pond bottom next to the lodge. The tops of these branches frozen into the pond ice can be seen at the right of the illustration; they are a sure sign of occupancy.

Beavers do not always build classic lodges. Where the soil is unsuitable, they often prefer to dig a bank burrow with an underwater entrance. Sometimes they cover the entrance with a mound of sticks and mud, as they would with a free-standing lodge, but sometimes they do not. This sort of habitation is often used in ponds that lack a shallow spot away from the shore upon which to build. They are also used along rivers where the current would wash away a stick lodge. In such areas, diagonally cut and debarked sticks washed up on the bank as well as cutting sign on streamside saplings may be the chief clues to the residency of this animal.

Dams

Beavers build dams to create a pond that in turn provides them with a relatively safe haven from which they can forage near its edge. As the favored trees close to the pond's margin give out, the dam is built higher to flood new areas, and feeding continues. If a natural feature such as a ledge or a man-made feature such as a bridge abutment is available, beavers will happily use it as the foundation for their damming. The dams are constructed of debarked sticks and mud, much as is the lodge. If the dam is in good repair, then the pond is probably occupied. However, even the best maintained dam is not waterproof. Some flow seeps out through the interlaced branches and mud, keeping the outlet brook flowing at a stable rate.

Scat

Beavers normally defecate in water so their scat is not often seen. Looking into the shallows near a dam with polarized glasses may show a collection of round yellowish "dumplings" that have collected in a depression on the bottom. Figure 2.1 C shows a couple of droppings raised from the base of a beaver lodge at Kezar Lake in western Maine. When the spring thaw floods forested areas, beavers take advantage by cutting down trees and saplings in the inundated woods. When the floods recede, their scats can be left high and dry where they rapidly disintegrate into little piles of sawdust often so far from water that they are not readily related to beaver.

Warning: Beaver scats, along with those of other animals, sometimes contain the giardia cyst. Using or bathing in water from a beaver pond or its outlet brook can make you sick.

Habitat

Beavers can be found coast-to-coast wherever water and edible growth are present. Although trees are their most conspicuous food, they also consume lesser plants: lily pads are a summer staple while laurel and highbush blueberry that crowd the edges of a pond are eaten as well. Hardwood trees are favored over softwoods and indeed a predominance of pines around an established pond is a sign of long tenancy. Beavers may not abandon a pond after favored feed trees in reach of their flooding give out, as long as other sources of food like lily pads and their starchy tubers are still available.

Once a pond is abandoned, the dam breaks and the pond bottom is exposed. Its mud, rich in minerals leached by inflow from the surrounding terrain, soon supports a lush growth of herbaceous plants and shrubs that condition the soil and provide shade until, in the process called "forest succession," trees once again begin to grow. This in turn sets the stage for the eventual return of the beaver and the completion of the cycle.

Trackard 3 – Moose

Moose are huge animals that traditionally inhabit the boreal forests and bogs of the far North. In recent decades in the East, however, they have been moving southward, apparently attracted by the wetlands created by the resurgence of beavers. Beaver ponds provide a convenient source for the sort of sodium-rich aquatic plants upon which moose prefer to graze. A bull moose in New England measures at the shoulders to the height of a man and supports an annual growth of huge palmate antlers that, after the conclusion of the fall rut, are occasionally found lying in the snow. Despite the size of these antlers, it is said that they are not found more often because mice and other small rodents begin gnawing them immediately, recycling the calcium and reducing them in short order.

Tracks

Moose are ungulates, that is, hoofed animals. Figure 3.1 A from a photograph taken in Brownfield, Maine, presents a shallow hoof print at actual size. Note that the pointed end is in the direction of travel, that is, the moose is moving toward the top of the card. Like others of its order, Artiodactyla, the moose's hoof is composed of two large cloves, which are actually its middle toes, and two "dewclaws" that protrude from either side of the foot about 2–3 inches behind the cloves. Because of their location these dewclaws usually fail to show in walking prints on firm surfaces. The dewclaws are actually the animal's outer and inner toes, which have diminished and receded up the foot in the process of evolution. Figure 3.2 C shows them in a print recorded in spring snow on Popple Mountain in northern New Hampshire. Figure 3.2 D is a deeper impression from the Swift River in which a characteristic rocker shows just behind the dewclaws, and the track appears short because the tips of the clove impressions are hidden under the snow.

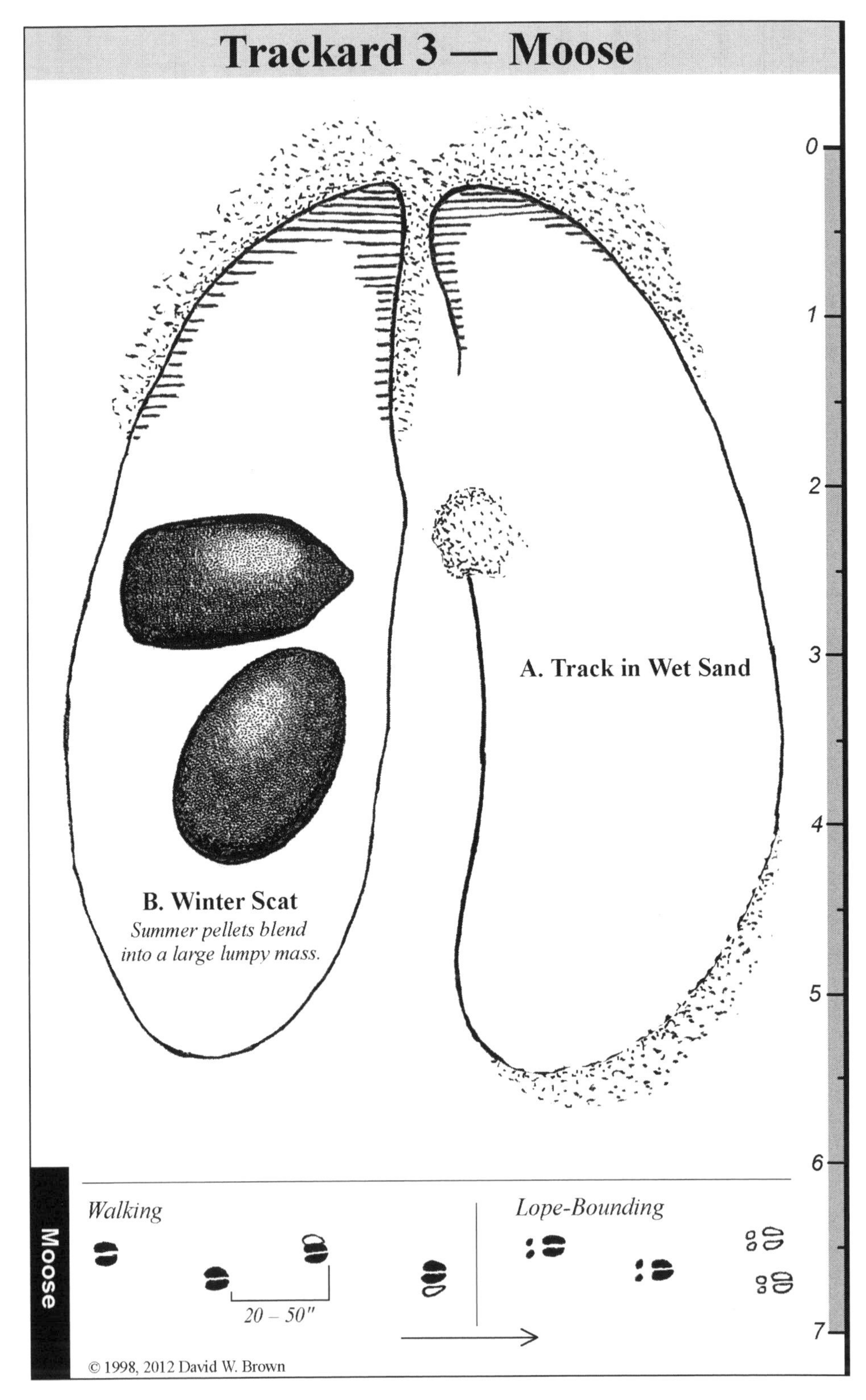

Figure 3.1. Trackard 3 – Moose.

In the East, where there are no other ungulates of its size, identification of adults is seldom difficult. However, a moose calf registering its prints in soft earth in early summer may easily be confused with an adult deer. Since a young calf will be with its mother upon whom it depends for protection, a look around may reveal her prints and solve the problem. Also deep prints in muck so saturated that water has collected in them are more likely to those of moose since these aquatic herbivores are often found in muddy bogs and seem to tolerate movement through such conditions better than deer.

Trails

The long-legged moose is an ungainly creature, whose walking prints often indirect and double register, especially in firm conditions. I have found lateral displacement to be marginally more common than with deer, that is, the hind print is impressed off-center from the front print and to the side. A typical walking trail, presented at the bottom of Trackard 3 (Figure 3.1), was recorded in spring corn snow on Wildcat Mountain in New Hampshire where the nightly freeze-up had permitted the animal to stay on the surface. When walking deer indirect register, on the other hand, the resulting trail more often shows an understep as opposed to a lateral displacement. The step length for walking moose in New England is commonly somewhere between 31 and 38 inches, but may be shorter or occasionally much longer; animals with long legs can stretch their steps quite a bit if there is reason to do so. The trail width of a walking moose ranges from about 10 to 18 inches.

Moose can also trot, lope and bound. These are excited gaits; either the moose is running away or running to challenge. Loping and bounding in snow or mud requires a great deal of energy for such large animals and so these gaits are not often seen. The example at the bottom right of Trackard 3 was from the deep snow of an abandoned lumber camp on Popple Mountain where the moose was forced out of the shallow snow of a conifer stand. The gait may have been caused by irritation with the deeper snow in the open and a desire to get across it as quickly as possible.

Spruce trees and many other conifers have horizontal dead branches on their lower trunks. Where moose move through stands of these trees, they snap off this "squaw wood" at the height of a man, creating a discernible tunnel through the forest. Like other hoofed animals, a habitually used moose trail tends to be lumpy underfoot

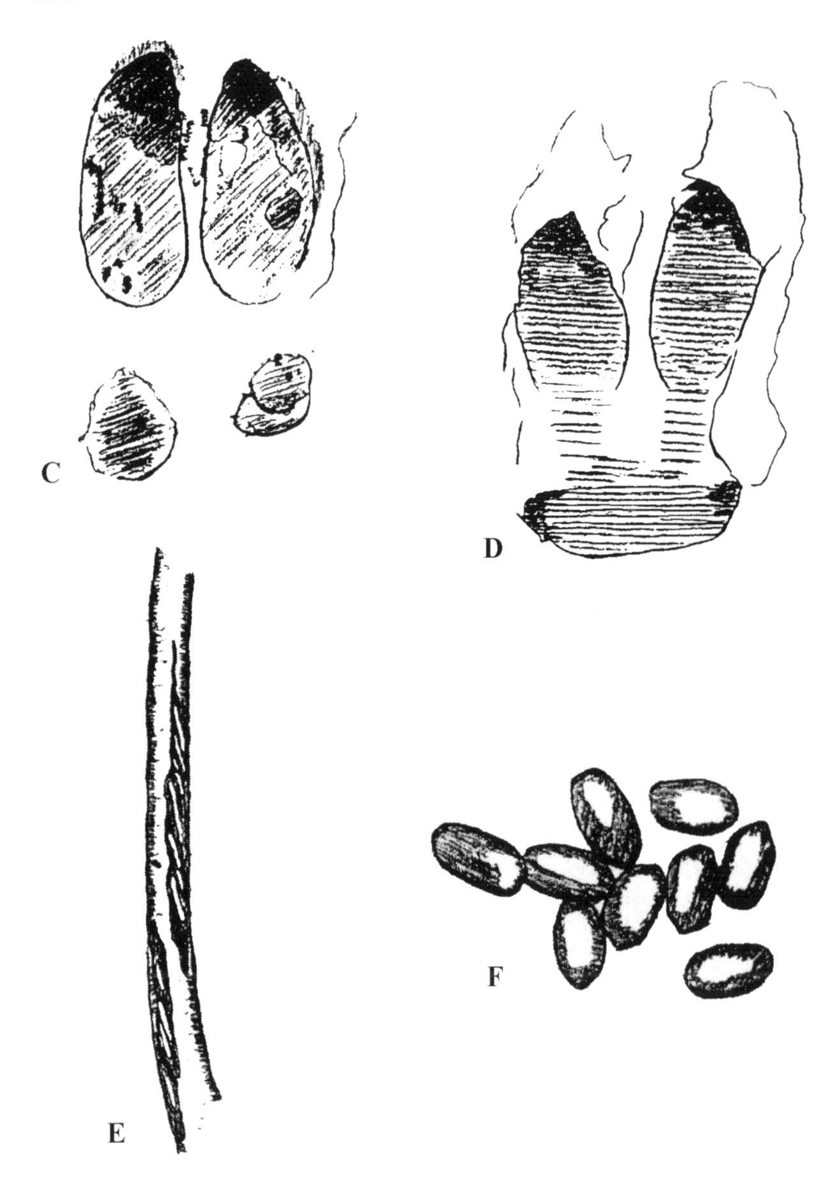

Figure 3.2. Moose tracks and sign, all reduced.

C. Print in snow showing dewclaws, about 1/5 size. Jackson, New Hampshire.

D. Print in deeper snow where dewclaws formed a characteristic rocker. Passaconaway, New Hampshire.

E. Bark gnaw on mountain ash. Jackson, New Hampshire.

F. Collection of winter scat, about 1/5 size. Jackson, New Hampshire.

compared to those of bears, which are padded smooth. Moose often use the machine-packed trails of cross-country ski centers to get from one stand of firs to the next. Their trails can be found all too easily in these locations as the animals posthole the carefully groomed ski tracks.

Moose often feed on the buds and twigs at the top of young trees. They may do this by stretching upward and grasping the tip with their mouths, pulling it downward and perhaps breaking it in the process. Or they may straddle the young sapling and force it down with their chests. Broken saplings an inch or two thick in regenerating clearcuts are a common sight in the North Country. (See the discussion of the differences in the appearances of breakage by bears, moose and porcupines in the Black Bear chapter.)

Scat

Figure 3.1 B shows typical winter moose droppings full size, and Figure 3.2 F a collection of the same from Popple Mountain. The scats are brown and are usually found in a cluster. Moose at this season live on dry feed such as bark, buds and fir twigs and so their scat forms into discreet pellets. In summer when moose are in the lowlands feeding on succulent aquatic vegetation in sloughs and beaver ponds, these pellets coalesce into a mass or "moose-pie." However, although stuck together, the individual pellets can usually be distinguished. Occasionally I find circular deposits of fine grain and even consistency that look like coffee grounds dumped out of an inverted fry pan. These seem to occur during seasonal transitions between dry and moist feed.

Sign Comparisons

Black bear: Moose often gnaw the bark of maples, mountain ash and other trees that have sweet cambium. The resulting gouges can be mistaken for the claw marks of a bear reaching up on the tree and dragging its claws down the trunk. When a bear does this, however, little tails of bark are left mostly at the bottom of the scar. Moose, like some other ungulates, do not have upper incisors. When they gnaw bark, they begin down low and scrape upward with their bottom teeth, leaving the tails at the top of the mark. Figure 3.2 E, from Popple Mountain, shows such scraping on a mountain ash. Anytime you see marks of this sort on a sweet-barked tree, you should first suspect moose rather than bear.

Deer: In winter moose often browse the twigs of young balsam firs, generally leaving the twigs of spruce alone. It has been suggested that they do this because their gut bacteria are adapted to the digestion of fir browse but not spruce. In view of the fact that moose also leave spiny plants like rubus (raspberries, blackberries and their relatives) alone suggests to me that the actual reason may be that their mouthparts are sensitive to the pointed needles of spruce, leading them to prefer the blunt needles of fir. Deer, on the other hand, readily browse spruce as well as rubus and greenbrier, so any browse found at appropriate height on these plants may generally be attributed to them rather than moose.

Moose browse can be told from deer in other ways, as well. Moose rip off thicker twigs than do deer and feed higher on the tree. Unless deer are starving, their browse line, the upper limit of their feeding, is at about the height of a man's chest. On the other hand, moose browse from knee-high to over a man's head. Willow wands from which the leaves have been stripped by running them sideways through the mouth are another moose browse sign. Indeed, any attention to willow more than a foot or two off the ground is usually attributable to moose as this is a staple browse for these animals.

When moose lie down in a bed, the resulting depression can be 4 feet wide and 5 feet long, or as long as 8 feet if nose or antler marks are present. By contrast, deer beds are around 2x3 feet. The moose's blackish hair is often found stuck to the compacted ice-snow in the bed, and dry winter scat of the sort illustrated will be found in the area. In the bed may also be found large yellow or sometimes reddish urine stains that have a pleasant herbal scent, to my nose a mixture of dill and freshly cut wood, which is similar to the common odor of deer urine when they, too, are feeding on woody browse.

Bulls in the fall "rut" or mating season, which begins in September, often rub certain young trees with their rack as a marking device. This is similar to the familiar "buck rubs" of deer. The rubbing shreds the bark and leaves tails at both the top and the bottom. But I tend to find that deer pick out a spindly young sapling only an inch or so in diameter whereas moose pick thicker saplings for this abuse. Note, however, that where antlered deer are not hunted out, bucks can reach very large size and may also pick thicker saplings for rubbing. Height above the ground is a better indicator: while buck rubs are normally between 1 and 3 feet off the ground, bull rubs center around 4–5 feet above it. In the West, the rubs of elk are at about the same height as those of moose and so other evidence must be found to more confidently identify the sign.

Habitat

In the warmer months moose favor the bottomland bogs of their northern range where they can stay cool in water and mud and feed on aquatic plants and other succulent vegetation. In winter, clearcuts are often visited for the buds on successional growth and stump sprouts that these areas provide. After beaver ponds freeze up in the early winter, moose frequently move upslope. In northern New Hampshire their winter destination is often the lower limit of boreal spruce/fir stands at about 2500 feet. At this elevation they have been observed using both the boreal forest and the northern hardwoods just below. Both forest types provide browsing and gnawing opportunities while the conifers also provide shallower snow, which collects in the dense branches before reaching the ground. It has also been advanced that a preference for such areas is a function of thermal regulation. Having large body mass, moose on mild winter days are in danger of overheating, while on cold winter nights, the evergreens that have been warmed during the day by the sun re-radiate their warmth to the animals that shelter within them. Such dense stands also provide protection from the convective cooling of wind. Thus moose at the lower border of the boreal forest may move into and out of hardwoods and spruce/fir stands according to their need to warm up or cool down.

Moose rut (mate) beginning in the early fall. During this time bulls are irascible and should be treated with respect by the tracker. Flee from a bull moose that moves toward you, head down and wagging its palmate antlers. As moose vision is not very acute, a sharp turn among the trees may throw the animal off your trail. Likewise a cow moose with a calf should also be regarded as dangerous even though the mother does not seem disturbed by your presence. An attack by either sex can come quite suddenly, while you are looking down to check the exposure on your camera, for instance. The lore of the North suggests that once a moose is upon you, it will stomp until you stop moving. My own close encounters with these animals suggests that, aside from the rutting period, their approach is more out of curiosity than aggression. Their near-sightedness forces them to come close to see what you are.

Trackard 4 – Snowshoe Hare

Snowshoe hares are named for their hind feet, which can spread to incredible width to support the weight of the animal on snow. These lagomorphs (rabbits and hares) are also called "varying hares" because in late fall their brown fur turns white for the coming winter. In the brown summer phase they are the same color as, and only slightly larger than, cottontail rabbits, whose northern range overlaps that of the hare, creating some confusion at least in appearance. Given the size difference of their feet, however, distinguishing the two by their tracks is easy.

Tracks

When pads can be discerned at all on their heavily furred feet, snowshoe hare prints usually show four toes on both the front and hind impressions. Sometimes a fifth impression can be detected on the medial side, retarded quite far back in the print. In prints with closed toes, the arrangement of these vague pads and their associated nails is asymmetrical. When the toes on the hind foot are spread, however, their arrangement becomes more balanced. Figure 4.1 A, adapted from photographs taken in various parts of the White Mountains of New Hampshire, demonstrates the amazing difference in size between a print in which the toes were close laterally and a print where they had been spread as wide as possible. The scale at the right edge of the card quantifies this dramatic difference. Partly because of this range of sizes, snowshoe hares, along with raccoons, are the great foolers, capable of making their prints look like those of many other animals. Snowshoe hare prints are often confused with much larger animals such as lynx, bear and mountain lion, while Figure 4.2 E, from the Passaconaway intervale in the White Mountains of New Hampshire, shows various prints that look rather canine. The points at the end of some of the toe impressions are not claws, however, as they would be

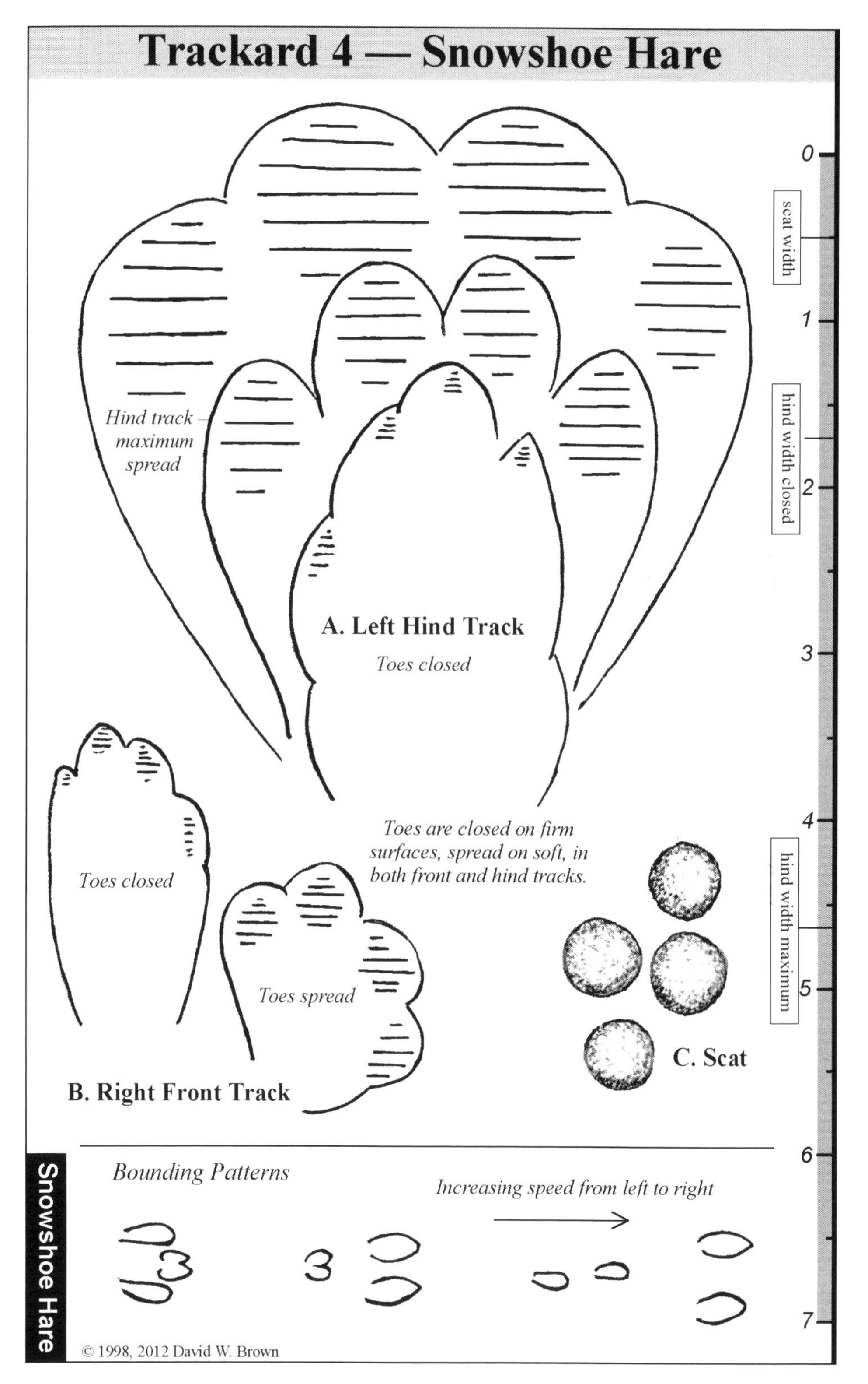

Figure 4.1. Trackard 4 – Snowshoe Hare.

in a dog print, but rather the tapered fur covering the animal's toes, drawn to a point like the end of an artist's paintbrush.

Snowshoe hares vary their print width in response to the amount of support that the snow surface provides. In Figure 4.2 F, from New Salem, Massachusetts, the hare spread its toes, seeking maximum support on fresh snow where the surface had been slightly stiffened by freezing drizzle. In other situations only the toes may show distinctly, radically changing the appearance of the print.

Because the feet are covered with dense fur, secondary pads register indistinctly or not at all. When they do, they are often just a series of dots such as those in Figure 4.2 F.

Front prints, which normally register to the rear in typical bounding patterns, are much smaller than the hind. However, the front feet have the same ability to expand, as demonstrated in Figure 4.1 B from Passaconaway. Hares can mix and match their toe-spreads to suit conditions. In slightly firmed snow, for instance, the hare may expand its front feet to support the weight of its head while closing its hind, thereby reducing the size difference.

Track Comparisons

Cottontail rabbit: Although cottontails have similarly shaped feet and show the same track patterns as do snowshoe hares, the maximum width of a cottontail print with spread hind toes is less than the minimum width for snowshoe hares with their hind toes laterally closed. As a "rule of thumb," cottontail hind prints (found at the head of each pattern) with toes closed are roughly the size of a man's thumb. Hare prints are much larger.

Trails

Confusion of snowshoe hares with other animals usually occurs where only one or two prints are available. If an entire four-print pattern (4X) is visible, identification is not difficult since hares, like other lagomorphs as well as squirrels, almost always use a characteristic bounding pattern, with the larger size of a hare's prints distinguishing its from those of the other species.

In the lagomorph version of this bounding pattern, illustrated in the trail at the bottom of Trackard 4 (Figure 4.1) from Gorham, New Hampshire, two hind prints are forward, even, parallel and wide apart while the two smaller front prints in the right hand pattern are retarded and nearly in line. Rabbits and hares have wide pelvises and very narrow chests, an anatomy that accounts for most of these pattern features. When front prints appear side-by-side, as in the two left hand patterns, they will be contiguous, again a feature of a narrow chest. The snowshoe hare's bounding pattern with its large size is so distinctive that it allows for positive identification even in poor conditions. Figure 4.2 D, from Passaconaway, with spread toes on breakable crust, shows such an easily identifiable pattern despite total lack of pad definition.

Within their home range, hares move around over well-established trails that gradually become worn down over time. In grassy or other habitats with soft ground these trails appear as shallow, meandering troughs about 6 inches wide. After a fresh snowfall, hares will begin padding down their trails by hop-bounding from one spot to the next. Gradually each landing spot enlarges into a "hare island," a characteristic chain of which is shown in Figure 4.2 H from Maple Mountain in New Hampshire. Eventually these islands will lengthen until they connect and become a continuous, packed trail.

Scat

Hare scat looks like that of rabbits, flattened spheres of yellow or brown grainy material such as in Figure 4.1 C, from Wildcat Mountain in New Hampshire. On average, snowshoe hare scat is slightly larger than that of cottontails. If a cluster of droppings is found, careful measurement should reveal whether at least one or two are as large as ½ inch in maximum diameter. If these are found and the habitat is open brush or dense, low evergreens, it is pretty safe to assume snowshoe hare. If most of the sample is slightly smaller and the area has thick deciduous brush, then Eastern cottontail is more likely. However, on mountainsides, where you would expect hares, another species of rabbit, the scarcer New England cottontail, may also be found. Samples of its scat tend to be small, more spherical, sometimes pointed, and, therefore, unlikely to be confused with hare scat. Of course young or small hares may leave cottontail-size scat, and in exceptionally good habitat, cottontails can consistently leave scat as large as that of hares. Tracking rules are seldom 100% applicable; they have to be imbedded in the circumstances in which the sign is found.

Hares also leave urine in the snow. Initially this deposit is often a bright orange dot that diffuses into the snow with time, leaving a surprisingly large pinkish area.

Figure 4.2. Snowshoe hare tracks and sign, all reduced.
D. Spread bounding pattern in breakable crust, greatly reduced.
E. Canine-like prints in snow. Passaconaway, New Hampshire.
F. Shallow hind prints on snow. The surface had been stiffened by drizzle.
G. Heavy browse, 3 feet off ground in snow country. Jackson, New Hampshire.
H. "Hare islands" in fresh, soft snow.

Sign

Like rodents, lagomorphs browse by making perfect diagonal cuts on twigs with their central incisors. Sometimes these cuts can be quite far up on a plant and appear to be the work of giant hare. However, such high cuts were probably made after a snowfall that raised the feeding level. Snowfall that deepens continually over the winter is very helpful to snowshoe hares since their huge hind feet enable them to stand on top of the soft surface to reach new browse. A snowless winter, on the other hand, is a great hardship for them, and many starve. When hares browse repeatedly on favorite shrubs, the leading twigs that are cut off are replaced by lateral leaders. When these, too, are nipped, more grow, eventually forming a bulbous clump such as the one shown in Figure 4.2 G from a photograph taken on Wildcat Mountain.

Judging by the amount of snowshoe hare fur in the scats of various predators, these animals are a staple prey species. Besides raising the browsing level, deep, soft snow also helps snowshoe hares survive this predation pressure. With their large hind feet, they are able not only to stand on, but also to run easily over a snow surface in which predators such as foxes and bobcats flounder. Only the fisher and the lynx, with their own snowshoe feet, are capable of catching them in these conditions. White hair found in the winter scat of these predators is usually from hares.

If the predator that left the scat had been scavenging a winter deer kill, however, there can be some confusion. While the white hairs of a deer from its back and flanks will have a dark tip, the coloration of their gray winter coat, and the tail hair is thick and wavy, the belly fur of a deer is quite fine and can be difficult to tell from that of hare. It is sometimes supposed that only deer hair will break at right angles when bent, but this is true of all white fur, which acquires its color not from white pigment that would fill its core and make it pliable, but rather to the reflection of light off its hollow surfaces. Being hollow, white fur, whether from a deer or a hare, breaks like a soda straw when bent. If fine white hair is found in a scat, other evidence must be sought by fractionating the deposit.

Warning: As will be mentioned frequently in this book, the scats of animals can contain breathable spores which can make you very sick. This is particularly the case with dry scat.

The hollow construction of white fur provides superior insulation compared to dark fur, the hairs of which are filled with solid melanin granules that make them more conductive. Obviously, as well, a white prey animal moving over the snow is less visible to the goshawk during the day and the great horned owl at night.

Most hare activity is nocturnal; in daylight these animals usually crouch under the boughs of young fir or spruce trees and wait for dusk. The approach of danger may elicit a single warning thump of a hind foot that will alert the careful observer to the animal's presence. Although a common mammal in its northern range, it is not seen often. Its initial instinct is to hold still, relying on its camouflage until danger has passed. If flushed, the hare will escape over its system of familiar, pre-packed trails.

Habitat

Snowshoe hares prefer forested areas with open brush or tree regeneration at browsable height and nearby low evergreens for concealment. In New England laurel thickets and conifer regeneration, especially if near blueberry patches, provide ideal habitat. Feeding areas may include brushy clearcuts, edges of forest roads and trails, old lumber camps, powerline and railroad cuts, streambanks, wetlands and so forth. It should be noted that hares are one of few animals that will browse the needles of pine seedlings, and sharp diagonal cuts on them at appropriate height can generally be attributed to this species. A look around will disclose scats deposited at random. The range of snowshoe hares is limited to northern areas with snowfall that makes the adaptations of large feet and white fur advantageous.

Trackard 5 – River Otter

River otters are large aquatic weasels adapted to preying on fish, crayfish and invertebrates, as well as on occasional small mammals. Their recovery in the East has tracked that of the beaver, which supplies them with watery habitat. Otters are social creatures, often seen in groups of two or three as they undulate across a pond in search of prey. Their native intelligence and easy life, with plenty of prey and few natural enemies, allow them the luxury of play, a characteristic unusual among adult wild animals, whose lives are generally desperate efforts to procreate and avoid starving or being preyed upon in the process. Otters love to slide and will take every opportunity to do so in either mud or snow, often repeating a downhill ride like a child with a sled. Any smooth surface will do, including crusted snow or pond ice. They have poor eyesight at least above water, a defect that they make up for with an acute sense of smell. During the day they alternate periods of hunting with rest and sorties ashore to excrete and eliminate. Otters can be vocal, uttering a variety of snorts, wuffs, chatters and chirps, especially when alarmed.

Tracks

Like those of other animals that show five toe pads, otter prints are asymmetrical, retarded on the medial side. This characteristic is more pronounced in the hind print, which is also slightly larger than the front. Both of these effects are clear in Figure 5.1 A and B, a set of mud prints from the Saugus River in Massachusetts. The medial fifth toe on the hind print often seems like a design afterthought, tacked onto the foot at the last moment of creation. Nails may or may not show. When they do they are close to the toe pads, often creating the teardrop shape typical of the weasel family.

Secondary pads register well back from the toe pads often as a series of connected circles, leaving a wavy outline both on the leading and rear edges. In many prints, the phalangials also appear, vaguely connecting each toe pad to a secondary pad protuberance. The soles of an otter's feet are furless at all seasons, so these bones often register.

All of the toes on both front and hind feet are connected by a web membrane that may show in a print, but its impression is not dependable. Its flexibility results in, at best, a light indentation that may be lost against a background of the same color and consistency as that of the print itself.

All five toes may not register in a print either. As with other weasels, the otter's medial toe sometimes fails to show. Figure 5.2 D, also recorded along the Saugus River, shows otter tracks in various levels of entirety, largely depending on their depth. Occasionally, a hind print will show a heel area, particularly in soft snow.

Trails

Several typical bounding patterns are illustrated at the bottom of Trackard 5 (Figure 5.1). Although they show the characteristic repeating slant arrangement of other weasels, otter patterns are often more ragged, with many indirect and double registrations. The deeper the prints, the neater will be the pattern. In firm conditions the patterns tend to spread out in the direction of travel, grading into the loping patterns shown in the middle section.

Otter trails are usually found near water. In the winter, these animals may use weaknesses in the ice near sunny banks or beaver dams in order to enter and leave a pond. Places where the ice on a beaver pond buckles due to changing water levels may also be used for access. Otters will travel surprisingly long distances overland from one body of water to another, crossing intervening hills, brush and woodland cliffs, locations more typical of fisher with whom their trails can be confused.

Track and Trail Comparisons

Fisher: Fisher tracks resemble those of otters in both size and shape. However, fur between the pads of a fisher's feet conceals the toe bones in their prints. Fishers also have front feet that are the same size or slightly larger than the hind, just the opposite of otters.

Overland otter trails are more or less direct routes between water bodies, while fishers circulate, investigating likely spots for hidden prey. Also, fishers do not enjoy sliding, while otters will always

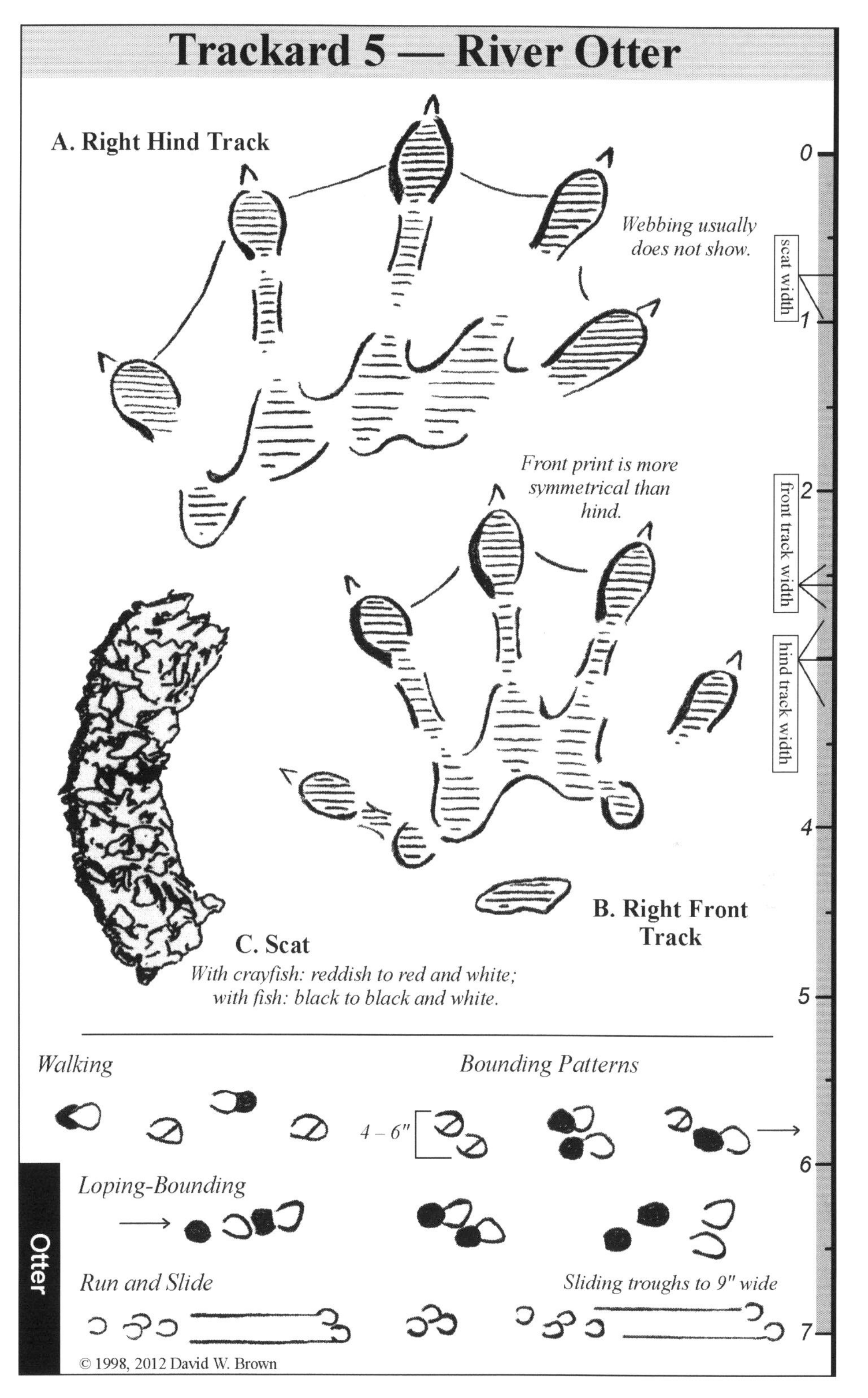

Figure 5.1. Trackard 5 – River Otter.

do so if they can, even on a level surface of firm snow or crust. This was the case in the trail at the extreme bottom of Trackard 5, recorded on the frozen surface of Quabbin Reservoir in central Massachusetts. Otter trails eventually enter water, while the winter trails of fishers do not. Fishers will drink from open water, however, so some care needs to be taken in investigating a trail that approaches an opening in the ice.

Raccoon: When a raccoon contracts its toes longitudinally and walks on them without heel contact, its prints can look superficially like those of a small otter. Note the raccoon signature, however: a smooth leading edge on a fused secondary pad compared to the lumpy and wavy-edged secondary pad of an otter.

Dog and **cat:** When the medial fifth toe does not register in a print and when the phalangials are withdrawn above the surface, the results can look surprisingly canine or feline as in the lower left print in Figure 5.2 D. Note, however, the large space between the otter's toes and secondary pads, contrasting with the crowded look especially of canid prints.

Beaver: Embankments repeatedly climbed by beavers will often develop a trough which may be mistaken for an otter slide, especially when viewed from an enforced distance, across an impassable stream perhaps, where small details like tracks cannot be discerned. However, beaver trails often switchback as the animal seeks the path of least effort to the top. On the way down, a beaver is usually dragging branches behind, the striated impressions of which will show in snow. In summer or fall, disturbed leaf cover or grass combed downslope will also betray this dragging activity of beavers. An otter slide, on the other hand, is more likely to be a narrow trough, 7–9 inches wide and will run down the fall line.

Porcupine: Porcupines, which are much smaller than beavers, often leave troughs in the snow. A look at Trackard 9, however, will show a rhythmically wavy trough compared to the otter's more or less straight, or at least randomly irregular, path.

Scat

Otter scat is normally composed of fish bones, fish scales or chiten from the exoskeletons of crayfish. Rarely, fur scat may be found, composed of the remains of young muskrats or beavers that the otter may have killed by invading an unattended lodge or bank burrow.

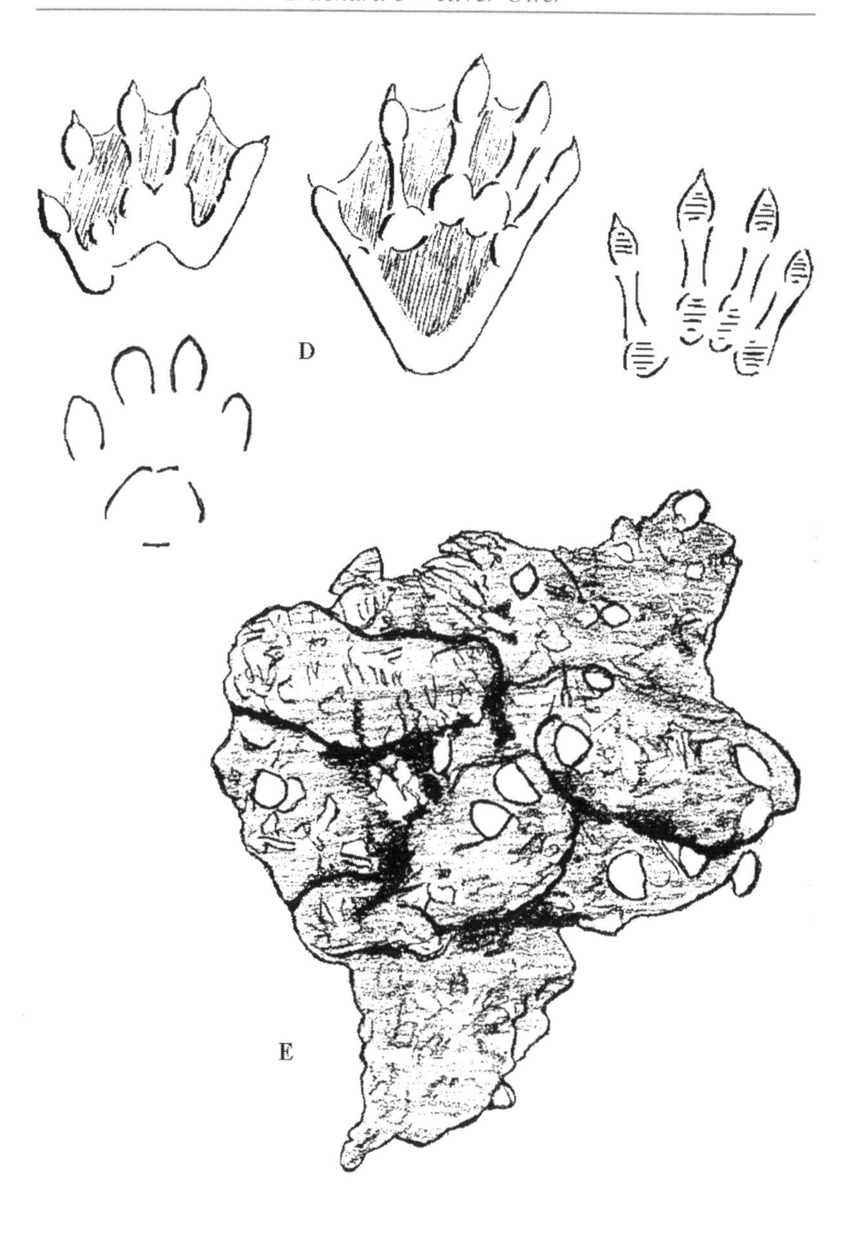

Figure 5.2. Otter tracks and scat.
D. Various tracks, about 1/5 size. Note the cat-like appearance of the four-toed print on the lower left. Saugus, Massachusetts.
E. Fish scat, actual size. Petersham, Massachusetts.

Otters spend most of their time in the water, coming ashore at habitual haul-outs, or "scat stations" to eliminate. There, scat and urine are deposited either on a raised mound such as a root hump or on a pile of vegetation scraped together as a substitute. Fish scat is usually a dark mass in which individual lumps may be discerned that average about ¾-inch maximum diameter. Such a deposit is shown in Figure 5.2 E, actual size. With time the soft content is consumed by bacteria or evaporates, leaving lumps of greenish white fish scales and finely chewed bones that disintegrate rapidly. An odor of fish oil will linger on the scat for some time. Droppings composed of crayfish remains also disintegrate quickly from reddish tubes to the red and white cracker-jack appearance in Figure 5.1 C, recorded on the Saugus River, and finally to a shapeless mass of chiten. A third sort of scat is composed of a runny, tarry mass, sometimes with yellowish content, that I have attributed to feeding on mussels, the shells of which are opened by puncturing them with the canine teeth.

A peculiar white mucous is sometimes found on top of otter scat. Some video that I recorded years ago shows this being expelled at the end of a scat deposition, and so I have assumed that it may be a lining for the intestines to prevent puncture by the sharp ends of fish bones. If it is rained upon, such a deposit may emulsify and last for some time at a haul-out as a pale, rubbery mass.

A typical haul-out is located near the edge of a beaver pond or streambank. Since otters revisit a favorite site, several deposits may be located within a few square yards. Around the mounds of vegetation may be found patches of pulverized conifer needles or forest duff where the animals have pushed their chests and bellies around on the ground, apparently for a pleasant scratch. When a site gets too full of scat to permit this, the otters switch to another site, returning to the first only after weathering and disintegration has cleaned it. Otters often seek out a peninsula in a pond or lake for these haul-outs, crossing from one side to the other, depositing scat and urine on the highest point between. These may be used so frequently that a path from one side to the other can be discerned.

Scat Comparisons

Raccoon: Raccoons feed on many of the same things that otters eat, are about the same size, and deposit scat along the margins of water, often repeatedly on the raised root hump of a tree. As a result, their scat, especially if it is composed of crayfish exoskeletons, is sometimes indistinguishable from that of otters. However, raccoons urinate at random, not necessarily waiting until they arrive at a defecation site. Otters, on the other hand, come ashore specifically to void, always urinating copiously on or around the scat. On bank moss, this often shows as distinctive large yellow or brown burn marks where the acidic fluid has killed the vegetation. Since raccoons are omnivorous, various lumps of scat at one of their scat stations may show different colors and consistencies representing their eclectic diet. Otter scat, or at least the solid sorts that are confusable with raccoon, is usually composed almost exclusively of fish or crayfish remains.

Herons: Herons often cough up a pellet of fish bones or crayfish shells at favored sites along water. However, they do not scrape together vegetation on which to deposit these droppings. A heron pellet will appear as a single discreet lump, deposited at random, with only one of the same age. Otter deposits are often in multiple segments. A look around may also disclose a large whitewash stain where the bird eliminated.

Habitat

Any watery habitat may be visited by otters, as long as it has fish or crayfish. The water need not be sparkling clean as is sometimes suggested. Murky beaver ponds, for instance, are often hunted thoroughly, the otters even using the host's lodge for dozing and sliding. Around civilization otters become wary and confine their hunting to hours when humans and their dogs are not likely to be around. Nevertheless, a familiarity with their tracks and sign will reveal their unexpectedly common presence.

Trackard 6 – Fisher

Fishers are large members of the weasel family, not cats as is sometimes thought. They are intermediate in size between the smaller mink and larger wolverine, with males as much as twice the size of females. Their fur is dull black or brown, often with some grizzling around the face and head. In other respects they have the usual weasel shape: long body and short legs. This anatomy allows them to attack prey in holes either in the ground, in snow or in trees. Fishers are thought of as arboreal hunters, but most of their time, especially that of the larger male, is spent on the ground where, like foxes and coyotes, they hunt by putting in a lot of miles. Snowshoe hares are a staple, but they often hunt squirrels and other small creatures as well. The male, at least, is able to kill animals as large and resourceful as a raccoon.

The deforestation of New England for agriculture led to near regional extinction of the fisher in the 18th Century. However, with the regrowth of the Eastern forest, fishers were reintroduced in the 1940s by forestry interests who hoped to control the porcupine population, animals that fishers are adept at killing. Since then the animal has spread out from its point of introduction in northern Vermont and from a possible relict population in the White Mountains of New Hampshire to woodlands all over the region and even into metropolitan parks. Yet it remains largely unfamiliar to the general public, so secretive is it on its home range. A familiarity with its tracks will reveal its presence as more common than is generally supposed.

The marten, often erroneously called "pine marten," is a related member of the weasel family, intermediate in size between the fisher and mink. Although its usual reddish coat contrasts easily with the dark brown fisher, their track impressions are very similar, so much so that distinguishing a large marten from a small female fisher can be very difficult. Marten ranges are restricted to boreal forests of the

North whereas fishers have extended theirs into the hardwoods farther south.

Tracks

Fishers have five toes on each foot. The medial fifth toe, like that of other weasels, is retarded and de-emphasized but usually appears in a print. The result is the asymmetrical shape presented in illustrations of both male and female tracks on Trackard 6 (Figure 6.1 A and B) from the Conway River in New Hampshire. Nail marks are short and often joined to the impression of the toe pads in a teardrop shape. The secondary pad is also asymmetrical, dominant on the lateral side of the foot while narrowing and partly withdrawn on the medial. Fishers are sexually dimorphic with the male's larger size reflected in an adult print about an inch wider than that of an adult female (Figure 6.1 A and B). The front and hind feet of either sex are about the same size, but the front, which carries the weight of the animal's head, often splays more radically and can appear substantially larger. The front print sometimes shows a small tertiary pad, as well, but this is not reliable.

Fishers can spread or close their toes quite a bit and so width measurements should be applied with the average spread of the illustrated front prints in mind. When fishers splay their toes widely for support, on a light snow crust for instance, they often show a distinctive "ragmop" effect demonstrated in Figure 6.2 G. On the same page are other illustrations of tracks recorded on various surfaces in central and northern New England.

Trails

Fishers are solitary and hunt on the move rather than from ambush. In the course of a day with uniform snow conditions a fisher may not vary its gait at all, leaving a monotonous trail of identical patterns. Where the surface is firm, fishers lope, leaving one or another of the patterns illustrated at the extreme bottom of Trackard 6. The speed of the gait represented by these patterns increases from left to right. On breakable crust, the lighter female can use a faster loping gait than the male, improving her chances of hunting success over his.

In soft snow where their feet sink in more than 3–4 inches, fishers switch to a bounding gait represented in the upper right of the two lines on Trackard 6 and shown with increasing speed from left to

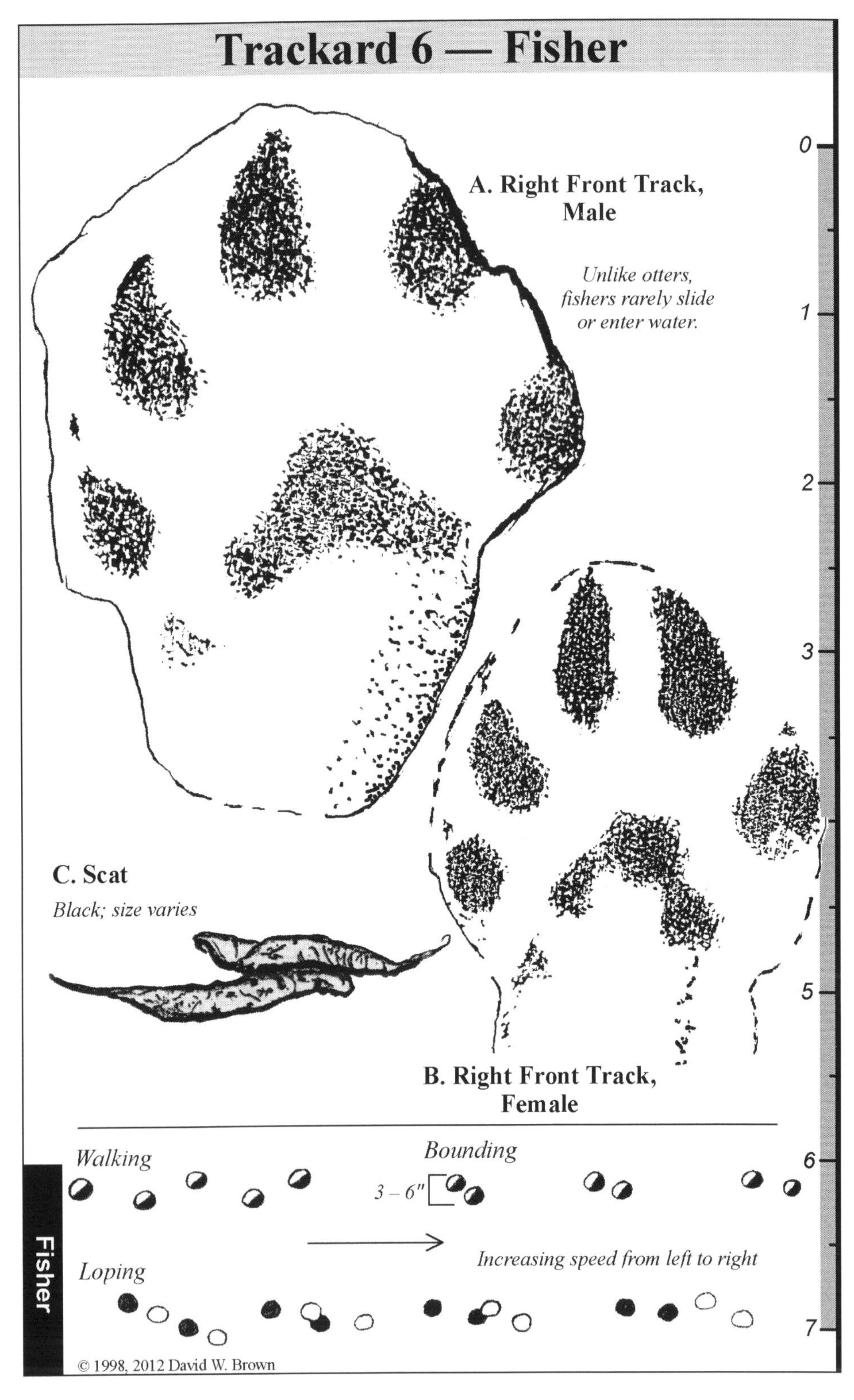

Figure 6.1. Trackard 6 – Fisher.

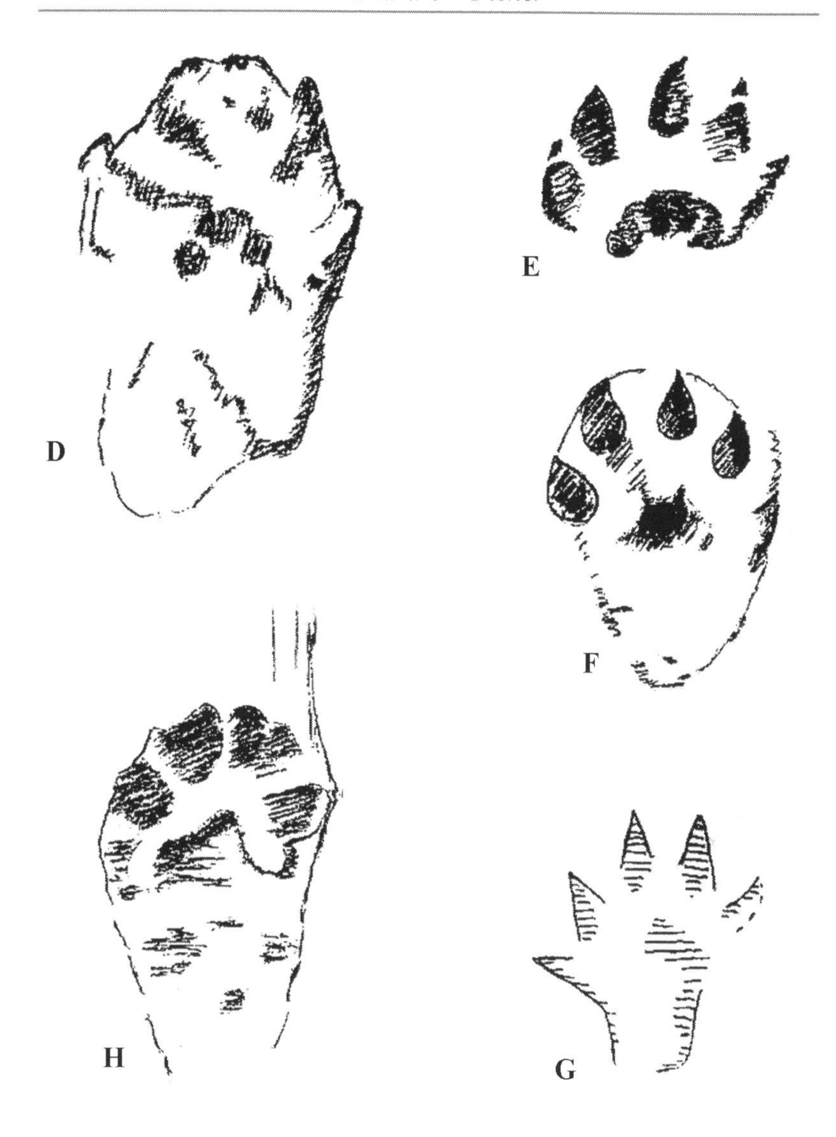

Figure 6.2. Various fisher tracks, about ½ size.

D. Left front of a male print in a skim of snow. Passaconaway, New Hampshire.

E. Print, ¾-inch deep, in damp, crusted snow. Princeton, Massachusetts.

F. Left front print, ¾-inch deep in soft snow over leaves. Conway, New Hampshire.

G. "Ragmop effect" in soft crust. Passaconaway, New Hampshire.

H. Print, 2-inch deep, in fresh, soft snow.

right. These are all direct registrations, with the hind feet landing in the prints vacated by the front feet. While a bounding gait requires high energy expenditure, the fisher makes up for this by pre-packing the soft snow with its front feet for the benefit of the hind, providing the hind with a firmer surface from which to push off into the next bound. The high arc of a bounding gait, which would be wasteful on a firm surface, lifts the animal clear of the intervening soft snow that a flatter loping gait would require the short-legged animal to plow through.

Fisher trails lead to holes in things. Although they are lightning fast for a few steps, their short legs are not well adapted to chasing down prey. They must either trap the quarry in a hole, chase it up a tree as with squirrels or surprise it dozing in brush as with rabbits. In winter, fishers hunt over a circular route that brings them back to the same features every few days. This practice provides two advantages. First, the fisher learns by repetition where prey animals on its circuit are likely to hide. In the intervening days between visits, those prey animals may forget about the fisher and make themselves more available on the predator's next pass. Secondly, in the event of lack of success, the fisher may return to caches along its route containing the remnants of carcasses from previous kills.

Track and Trail Comparisons

Otter: Otter prints are about the same size as those of fishers. However the front print of an otter is noticeably smaller than its hind while the front print of a fisher will register as the same size or somewhat larger than the hind. Phalangials will not appear in the print of the fisher, which has a lot of fur between its pads.

When otters travel overland between ponds or streams, their trails may be found well away from water and may be confused with those of fisher. Conversely fishers often hunt wetlands where otter trails may also be encountered. In addition to the differences in tracks appearance, certain characteristics of the trails of each can be relied upon to distinguish the two. First of all, fishers will not enter water although they will drink from its edge, so some care in interpreting a trail that comes and goes from the edge of water may be needed. Secondly, fishers hate to slide, only occasionally doing so, to get under a low branch for instance. Otters, of course, love this behavior, often loping and sliding even on the level surface of a snow-covered pond or brook. Following a perplexing trail in snow to a downslope will immediately tell the difference between the two.

Raccoon: A raccoon in certain conditions can leave a print very similar to a small or female fisher, especially with its front foot. However, the leading edge of a raccoon's secondary pad, both front and hind, will describe a smooth convex arc quite unlike the wavy leading edge of a fisher's. If a track pattern is available, the differences described in the following paragraph should be apparent.

Walking or trotting raccoons leave a two-print pattern, which might be confused with that of a small bounding fisher. However, the fisher's pattern will show tracks arranged on repeating slants while those of a raccoon will either be even or show alternating slants. Furthermore, the larger hind print of the raccoon will alternate sides of its trail with each stride, while the left and right direct registrations of a fisher will look identical.

Scat

Since fishers are primarily carnivores, their scat is usually composed of the remains of animal prey. The appearance of fisher scat can vary widely, however. Figure 6.1 C shows what is often represented as classic fisher scat. This deposit had been placed seemingly at random on December snow crust along the hunting route of the animal. Its small size belies the fact that it was excreted by a male. Figure 6.3 I-L show a number of scats of various sizes and shapes from various places in New England. Figure 6.3 K was found at a raccoon carcass next to which a fisher had denned for a couple of days. It is unusually large for a fisher, something I attribute to large meals at the raccoon's expense. However, it is also possible that this scat is an example of counter-marking by another predator attracted to the site by the scent of the kill. The scat in Figure 6.3 I was located at mid-winter on a snow-covered stump, a favorite location for fishers to mark their passage or their territory with a dropping. Fishers seem to be able forcibly to excrete small amounts of scat for this purpose. Figure 6.3 M, the smallest example shown, which was found on the butt end of a downed tree, is an example of this kind of deposit. Its size might easily cause it to be confused with that of a mink, especially since it was found in a wetland.

A well-known taste for berries may result in a scat quite different from the usual carnivore-type droppings. The loose deposit shown in

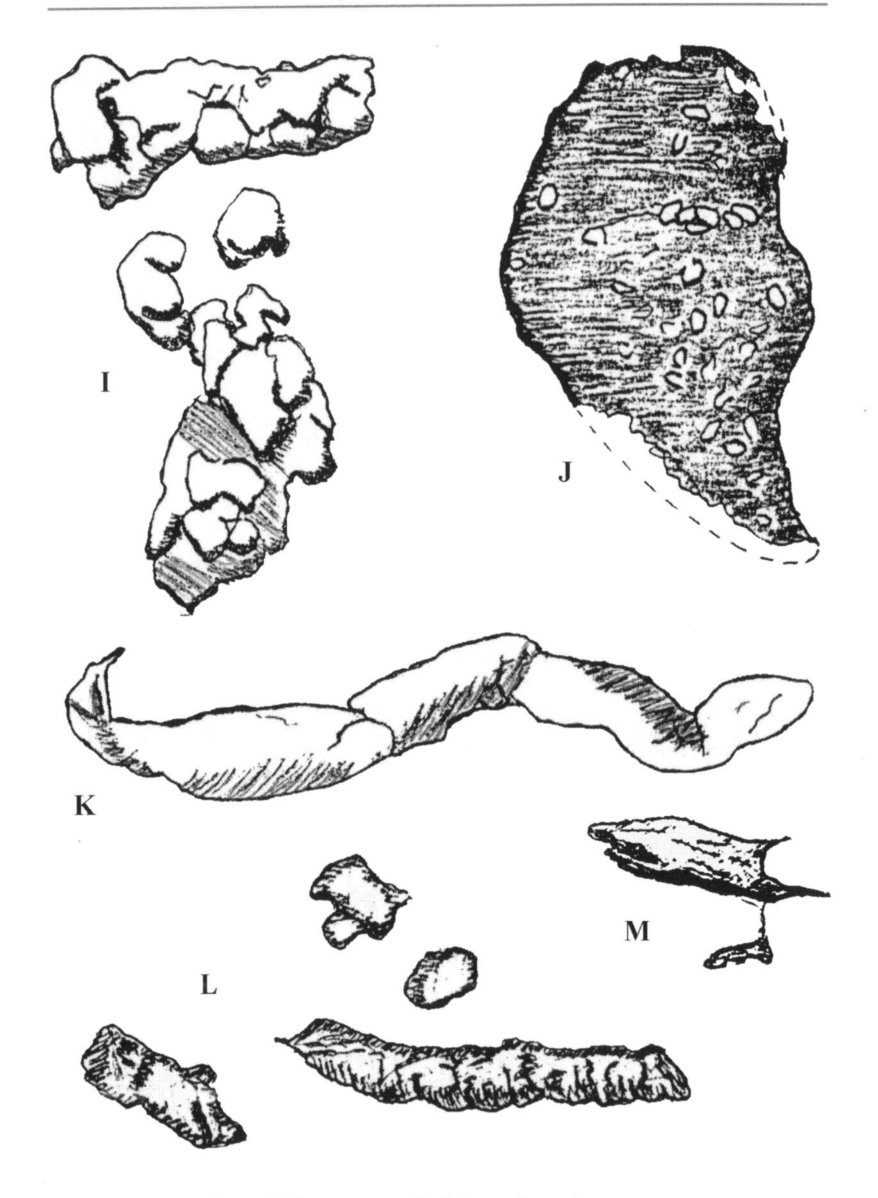

Figure 6.3. Various fisher scats, slightly reduced.

I. Crumbly tan-colored scat from stump top, winter. Conway, New Hampshire.

J. Brown berry scat on November snow. Carlisle, Massachusetts.

K. Yellowish hair scat from winter raccoon kill. Conway, New Hampshire.

L. Black winter scat from snow-covered stump. Jackson, New Hampshire.

M. Dark brown "force scat" on snow-covered fallen log. Conway, New Hampshire.

Figure 6.3 J was photographed in late November in Carlisle, Massachusetts, where the animal had apparently been eating berries remaining on bushes into early winter.

Fishers often hunt the same habitats as foxes and for the same prey. As they are of similar size, their scats can easily be confused. However, red foxes often leave scat along human paths, a habit not shared by fishers, which normally cross rather than follow human-created features. But in the absence of tracks or other indications it is risky to pin the identification of fisher scat on appearance or placement alone.

I have not often found fisher urine with a musky smell, as might be expected of a member of the weasel family with musk glands. It is normally pale and has only a faint odor, like the smell of chemically treated paperboard, an odor shared unfortunately with gray foxes, which also can leave fisher-size scats.

Sign

Fishers often visit rotting tree hulks in their ranges, tearing them apart like miniature bears in search of beetles, larvae, voles, mice or other small creatures hiding in the interior or in the softened root structure at the base. Fisher trails sometimes lead to trees with raptor or crow nests in them. By caching prey in such a nest, after it has been abandoned for the season by its owners, the fisher can conceal its food and raise it above the scent plane of scavengers like foxes and raccoons. In suburban parklands, such nests also provide the fisher with a resting platform during the day where it is unlikely to be disturbed by the many dogs that are brought to such woodlands.

Dens

A fisher may den for a while next to a cache or carcass, using any cavity available, such as a woodchuck burrow, the underside of a downed tree or a hollow in a brush pile. Long-term use of a den is limited to the female, who uses holes in trees for birthing and raising offspring. Around April the females, who with their lighter weight seem to climb more than males anyway, begin getting double reward for their trips off the ground; their investigations into cavities may result in finding both prey concealed within and potential birthing locations.

Habitat

Fishers can adapt to almost any forest type that contains abundant preferred prey. In the North Country they may concentrate on snowshoe hares and spend much time hunting in the dense, low conifer growth that this prey species prefers. However, as fishers have extended their ranges southward, they have adapted to other available prey. In the old oak woods of suburban parks in southern New England, for instance, they have developed the habit of hunting gray squirrels, which are abundant in such places.

For breeding, fishers need forests with at least some trees old and large enough to have developed cavities. These holes may result from the falling of limbs from a tree with rotted heartwood or from the work of a large woodpecker like a pileated, a species that has also proliferated with the rematuring of the Eastern forest. Thus fisher ranges may include both deciduous hardwoods and conifers, as long as there are denning locations and available prey. For this reason, old woods where standing and fallen deadwood is left in place are particularly inviting.

Trackard 7 – White-tailed Deer

Several species of deer inhabit North America, all yielding very similar tracks and sign. The two main divisions are between the mule deer and the white-tails, whose ranges overlap in the West along the eastern slope of the Rocky Mountains. Although there are subtle differences in track shape, the mule deer often showing a narrower hind print than the white-tail, discrimination on this characteristic alone is unreliable. In parts of the West both species also overlap with the much larger elk, which inhabits high forest habitats. Elk, too, have similar feet so that a young elk's tracks may easily be confused with either of the two smaller deer. As a general rule, mule deer prints tend to be slender and elk robust, with white-tails in between. Habitat may help, as well. In the Southwest I have found elk only in sub-alpine environments, while white-tails seem to prefer dense growth near water. In arid scrublands, on the other hand, I have only found mule deer.

In the East, deer were nearly eliminated by the start of the 18th Century as a result of over-hunting and land use patterns imported with colonists from Europe. Since the decline of farming east of the Appalachians after the Civil War, a fragmented forest has been growing back, providing deer with refuge in its interior and browse along its brushy edges. As a result of this ideal habitat as well as game management efforts, white-tailed deer numbers have exploded in many parts of the eastern seaboard. These animals normally browse buds and leafage from knee level up to the height of a man's chest, and in areas where there is a big herd, a browse-line under evergreens at about 5 feet off the ground is often visible. Deer are often seen grazing in fields, as well, especially in the spring before grass becomes tough and abrasive. At other times they may be cropping succulent plants that grow with the grass rather than the grass itself.

While the regrowth of the forest has aided the recovery of deer, their resurgence has also tracked that of beavers whose pond openings in the forest are edged with browsable plants and shrubbery of appropriate height. The discovery by deer that suburban rose bushes and other ornamental shrubs are nutritious and grow in areas where hunting is not permitted has drawn them closer to humans, sometimes straining the image of these beautiful animals with what were initially sympathetic suburbanites.

Tracks

A deer's foot is quite long; it extends from the tip of the cloves all the way up to what we tend to think of as the animal's elbow. The features that always register in mud or snow are the two central cloves such as those represented in Figure 7.1 A and B. These are actually the middle toes, while the "dewclaws," two discs situated behind the cloves, are vestigial lateral and medial toes, which have migrated back up the foot in the course of evolution. The dewclaws generally register only when the animal is moving at speed or in deep prints as was the case in Figure 7.1 B from Petersham, Massachusetts, in 5 inches of soft snow. The sharp ends of the cloves point in the direction of travel. Their leading edge is hard, somewhat like a human fingernail, while the interior of the clove and its rear margin have a softer, rubbery consistency. In deep tracks the tips of the cloves are often hidden under snow or earth, making the imprint seem shorter than in shallow prints. This effect is also shown in Figure 7.1 B as well as in Figure 7.2 D. The posterior of the clove is not as sharply defined as it appears in 2-dimensional illustrations but rather slopes away upward often leaving a vague outline. Both of these effects make it unwise to gauge the size of a deer from the length of its print. On the hard surface of a dirt road, for instance, only the tip of the cloves may register, making an adult deer's track look like that of a fawn. The central cloves are under muscular control like the toes of any other mammal and may be splayed widely when the deer needs support on muddy ground or at speed. Figure 7.2 E shows such splaying by a bounding deer.

The dewclaws are attached under the hide to tendons that provide for muscular control. They may be slanted relative to the direction of motion or arranged nearly parallel or perpendicular to it depending on the gait and the needs of the animal. In a normal walking gait with shallow track depth these dewclaws do not register, but when the deer is loping or bounding, they do. A deer's foot is collapsible to a certain extent, a feature that allows stretching tendons to absorb the

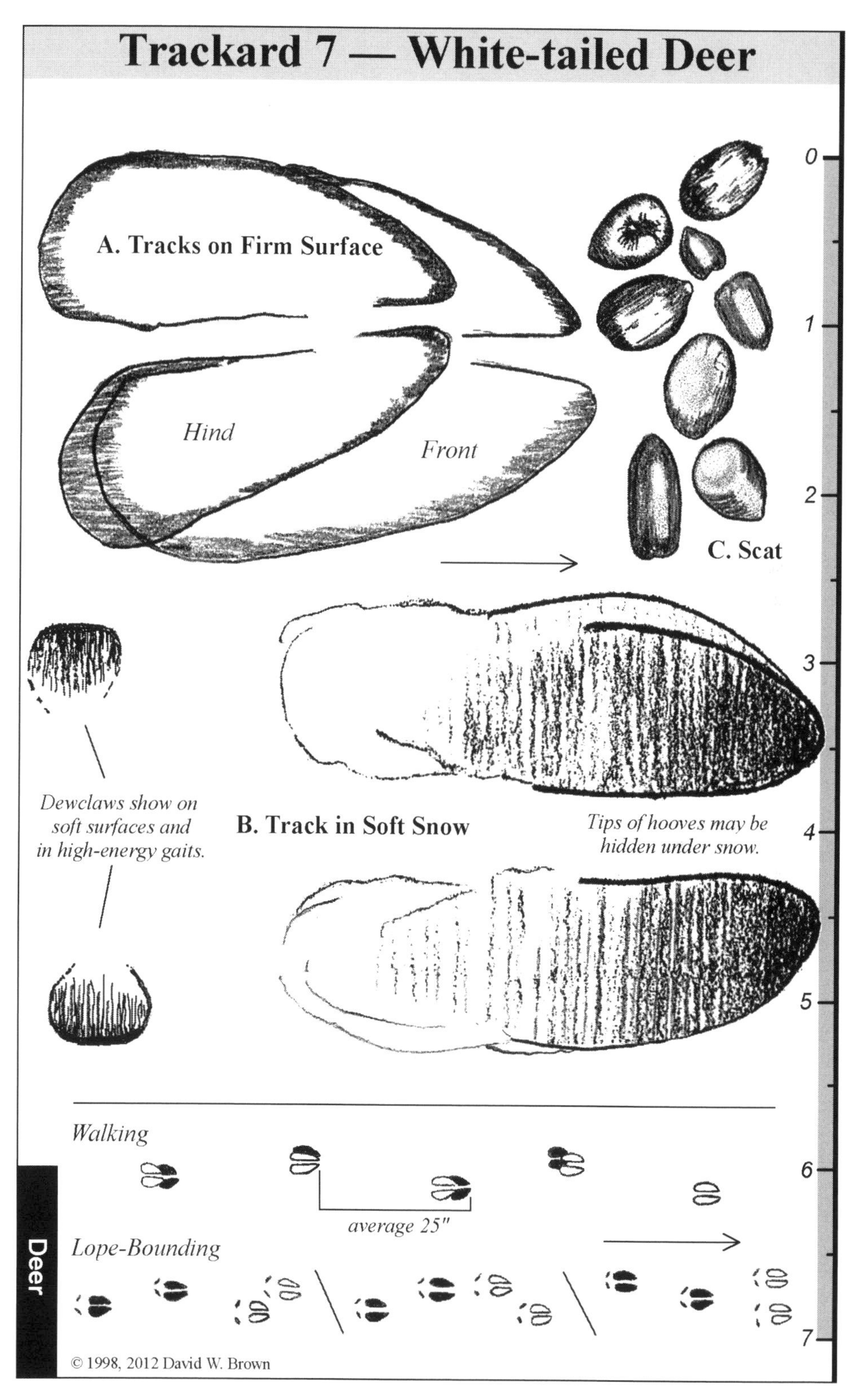

Figure 7.1. Trackard 7 – White-tailed Deer.

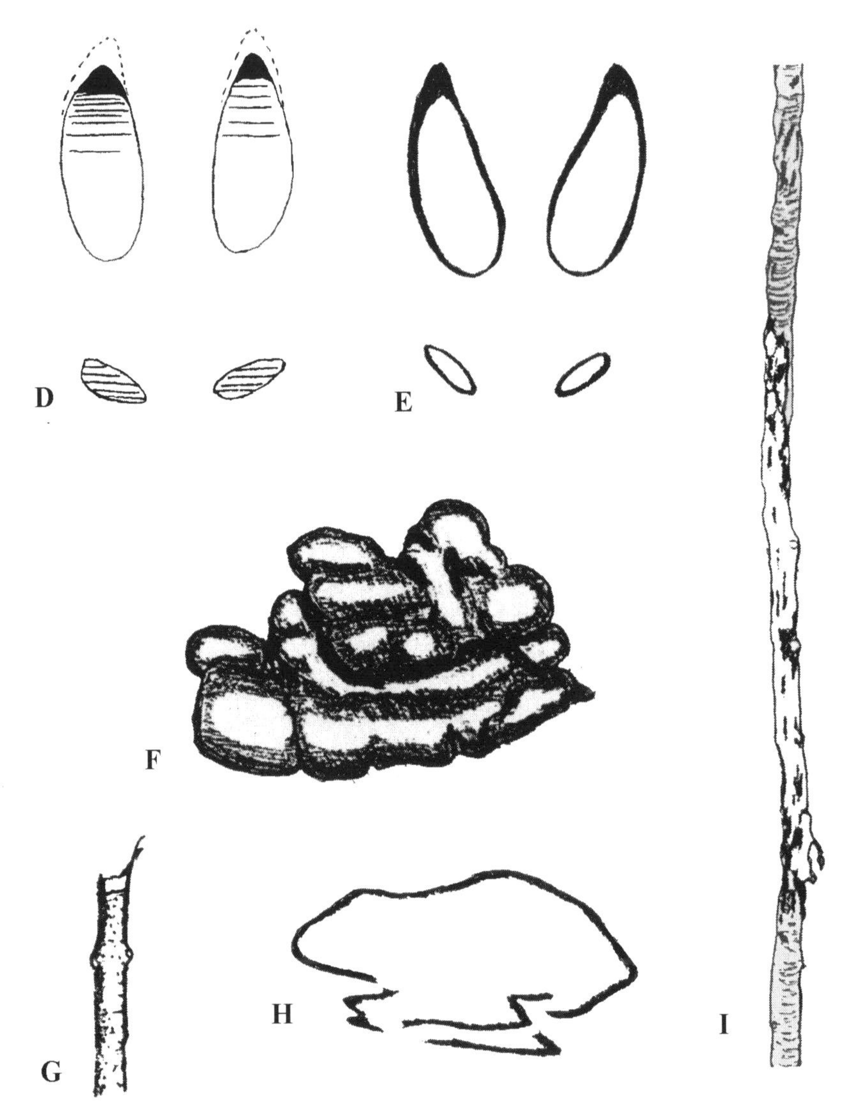

Figure 7.2. White-tailed deer tracks and sign, all reduced.

D. Track in deep snow where dewclaws have registered and tips of cloves are hidden under surface.

E. Typical bounding track with splayed cloves and dewclaws.

F. Summer scat, about ½ size. Concord, Massachusetts.

G. Typical browse.

H. Bed in thin snow. Petersham, Massachusetts.

I. Buck rub on narrow sapling, 1-inch in diameter. Concord, Massachusetts.

considerable energy of the animal's descending bulk at the end of each bound. As the collapse occurs, the dewclaws come into contact with the ground and appear in the tracks.

The medial fifth toe is vestigial and has migrated so far up the deer's foot that it never registers in prints. Since the heel area (the "elbow") of a deer only touches the ground when the animal is lying down, it is not padded.

It is also unwise to assume the sex of a deer by the size of its prints. In most parts of the country bucks are hunted more intensively than does with the result that few bucks get to grow to their full potential, while a doe may survive many hunting seasons and become larger than most bucks. The presence of fawn prints nearby is the best evidence for a doe.

Trails

Walking deer leave a trail such as the one at the bottom of Trackard 7 (Figure 7.1). Either the hind print is perfectly superimposed over the front or, more usually, it registers a little behind in a slight understep such as shown in Figure 7.1 A, recorded in wet sandy mud in New Salem, Massachusetts. Walking deer can also indirect register laterally, that is, the hind print may be offset slightly to one side of the front print, but I have found this sort of displacement more usual among moose.

Deer can live in a surprisingly small area that shifts with the seasonal availability of food. For most of the year they associate in small "herds," loose groups composed of either bucks of varying ages or does and yearlings including prepubescent males. In snow the members of these herds follow one another from bedding to feeding areas and back, packing down a lumpy trail that provides a route requiring less energy to walk than if each animal went its own way. As snow deepens, and especially if it crusts, the movement of deer becomes more restricted, and the packed down trails become circumscribed into a "yard," a series of interconnecting trails that often wind through evergreens where the animals can feed on needles, find protection from wind and the deeper snow elsewhere, and bed under conifer boughs that absorb warmth during the day and block radiational cooling at night.

Deer often drag their feet even in shallow snow, leaving a trough forward of each track or, if a slight stiffening of the surface has occurred, two lines representing the tips of the cloves. The walk of a deer in the winter is methodical, with no waste of effort or precious energy. Deer drag their feet in snow because it is easier than lifting them higher, not because they are incapable of doing so as is sometimes suggested.

Deer follow each other through the woods in non-snow seasons as well, leaving vague trails of packed down leaves. Such trails are easy to mistake for human paths. To the hand or even through soft-soled shoes, however, deer trails feel lumpy, especially if they are frozen; human footpaths tend to get padded down smooth like bear or porcupine trails. When deer trails pass over a low-point in a stone wall, the scraping of the sharp leading edges of cloven hooves is often evident on the stones. If the long-legged deer have to step over a fallen log along a path, they often make contact with it enough so that with time its upper surface gets abraded, another clue to the animal's presence.

Deer also lope and bound. Some of the resulting patterns are shown at the extreme bottom of Trackard 7. These gaits are generally reserved for flight from danger. In very deep, soft snow a bounding deer leaves a series of irregular rectangles in the snow.

Track and Trail Comparisons

Eastern coyote: One might not think that the tracks of these two vastly different species could be confused, but a deer's average print size is similar to that of an Eastern coyote, so much so that in a trail with vague impressions such as in melted and refrozen corn snow, they can be hard to distinguish. Even the two central nails of a coyote usually appear just where the points of a deer's cloves should be, while the other two claws, which might be counted on to identify a canid, often don't register in coyote prints. Several distinguishing features should be looked for. First, hoofed animals are the only ones that show "cliffing," that is, sharp angles between vertical and horizontal at the leading edge of their prints. The toe pads of other mammals are soft, with curved edges like the bottom of a bathtub and so tend to leave impressions with softer edges. A second difference between a coyote print and that of a deer may be found by running a finger along the leading edge of each toe impression. The clove area of a white-tailed deer's melted-out print will reveal a smooth outward curve while the same area of a vague coyote print will often feel flat or even slightly concave. Mule deer, on the other hand, may show this same concavity in nearly the same location. A third distinguishing feature is the sharp longitudinal ridge in the center of a deer's print. It often survives melting as a linear mound and contrasts with the rounder mass in the middle of old coyote prints in similar condition.

Although the step length of a walking deer, about 2 feet, is close to that of a walking Eastern coyote, the trail width averages a couple to several inches wider. Typically a coyote's trail looks like a nearly straight line of tracks while a deer's looks distinctly zigzagged.

Moose: A moose calf can leave a print like that of an adult deer. Unlike fawns, however, which are often left hidden by the doe while she feeds, moose calves stay with their mothers at all times for protection, so a look around may reveal her unmistakable tracks or droppings as well.

Scat

In winter, deer feed on dry browse such as the tips of hemlock twigs and other conifers as well as the buds of hardwood branches within reach. Figure 7.1 C, mostly from Petersham, shows variations in size and shape resulting from this diet. The pellets are discreet and can vary in size from so small as to be mistaken for squirrel to nearly the size of a moose calf's.

In warmer seasons when deer are feeding on more succulent vegetation, their scat often resembles Figure 7.2 F, photographed in Concord, Massachusetts. Such a clump is composed of pellets loosely stuck together.

Scat Comparisons

Rabbit and hare: While rabbit and hare droppings describe round or flattened spheres, winter deer scat has a more elongated shape with the pellets often showing a point as well as an indentation at one end or along a side. While cottontails sometimes show a similar point, the pellets are never elongated or indented. Deer scat is usually found in a pile where the animal was bedding, walking or feeding. Rabbit and hare scat is found as either individual items scattered around or as larger clusters near cutting sign typical for lagomorphs.

Moose: Moose calf scats look like miniature versions of the adult scat. While both deer and moose scat can show a point at one end, moose pellets tend to show indentations less often. Furthermore, a calf's scat will usually be found near the unmistakable deposits of its mother.

In the winter when their yellow urine marks are most easily detected against the snow, deer are feeding on the woodier vegetation of hemlock, yew and other evergreens. Urine deposits at this season have a pleasant herbal odor similar to dill. Moose do more bark gnawing at this season and so to their urine may be added the faint odor of "woodwine," a scent sometimes found in bourbon and furniture polish. However, the distinction is often slight at best.

During the rut in early winter bucks urinate on their legs in order to carry scent from a gland on their heel to the ground as an attractant for does. This sort of deposit has a harsher odor than the herbal smell of other deer urine. It is usually detected as a small yellow mark around the dewclaw area of the buck's print.

Sign

Deer lack upper incisors. When they browse a shrub, they press the twigs against their hard upper palate with their lower front teeth and jerk their head, tearing off the foliage as much as cutting it. The resulting twig end is shown in Figure 7.2 G. Note the rough perpendicular cut, often with a tag end. Rodents and lagomorphs make smoother cuts at 45 degrees to the axis of the twig.

Deer will browse the buds of deciduous limbs brought to the ground by winter ice and wind storms or by bears or porcupines. They may even venture out over the ice to visit the caches of beavers for the twigs that stick up above the ice outside the lodge, a dangerous practice should they be caught in the act by their predators. Lacking any of the above sources, deer will subsist on the needles of evergreens like hemlock even though they get little nutrition from them. Look for the characteristic frayed square-cut on outer twigs. This sign found either on pine or found higher than about 5 feet on any growth indicates deer in desperate straits as in late winter or early spring before "greening up."

During the rut, in the fall and early winter, bucks leave a variety of sign to attract does. When their antlers mature, they rub off the "velvet" by thrashing a bush or low branch on a tree, leaving damage behind along with festoons of bloody velvet that soon dry and disintegrate or are consumed by small rodents. Along trails frequented by does, the buck will scrape up a patch of dirt, urinate in it, and often snap an overhanging branch that then hangs down, as if pointing to the spot. In the Northeast these "buck scrapes" are often found under hemlock boughs.

Bucks also commonly rub their antlers on thin saplings along their trails. The trees selected have smooth, branchless bark on their lower portions; in New England poplars and sumac are often chosen. The buck lowers his head and inserts his rack so that the sapling is between its central tines. Then he rubs the base of one or both of his antlers against the bark, creating the wound illustrated in Figure 7.2 I. I find such rubs on saplings up to about 2 inches in diameter, although somewhat larger trees are reportedly used in other parts of the country, and from about a foot off the ground to waist high. Note that the frayed tails of inner bark will appear at both ends of the rub since the buck moves its antlers up and down. Bark gnawing by deer, on the other hand, tends to show most fraying at the top end of the scar where the bark was torn off.

Sign Comparisons

Moose: Moose usually nip off twigs substantially thicker than those of deer. Also moose seem to avoid sharply pointed browse such as spruce, greenbrier and rubus which deer regularly browse. Moose, of course, also browse higher than deer, although starving deer in late winter may be forced to stand on their hind legs to reach fresh foliage.

Bulls in the fall rut will rub saplings with their antlers just like bucks. However, moose generally choose thicker trees for rubbing and the torn bark will appear higher on the tree, from 3–4 feet at the bottom to about 6 feet at the top.

Squirrels: Deer will eat twigs and buds in the winter, but they will also paw up large patches of snow searching for acorns neglected by squirrels in the fall. Lacking the squirrel's nimble toes with which to manipulate the nut during feeding, deer simply crunch the acorns with their teeth, extracting as much meat as they can and leaving the rest. Squirrels, on the other hand, systematically peel the acorn shell to extract all the meat. Sometimes they will discard a partially opened nut which has either spoiled or has too high a tannin content, but a look at other acorns at the feeding site is sure to show the difference.

Porcupine: Porcupines also paw up large areas of snow after acorns and their feeding is incompetent enough to resemble that of deer. Look for tracks in any remaining snow or in the duff exposed and softened by the clearing action. The cliffing and ridging of deer prints compared to the rounder edges of porcupine prints should tell the difference in even the vaguest medium, although this may require you to track with your hands as much as your eyes. Also, when a deer is clearing snow off fallen leaves in search of acorns or other nuts, it performs a diagonal sweeping motion with one of its front feet at a time. Sometimes this results in a distinctive triangular pattern to the disturbance, often with a print of a front hoof in the center of the triangle. Thrown debris, such as dirt sprayed on the dead leaf background, also suggest the sweeping motion of the long-legged deer.

Debarking by either porcupines or other rodents may be mistaken for buck rubs, especially if this feeding was done while the animal was standing on a foot of snow or a low branch, elevating the wound on the tree. Close examination, however, will show neat gnawing with incisor marks where a buck rub will be smooth with rough edges.

Bear: Like porcupines, bears tear up large areas of leaves without apparent pattern. Acorn shells tend to be crunched and then swallowed, however, rather than being crunched and discarded as with deer and porcupines. Wild turkey flocks also tear up patches of dead leaves searching for acorns. Look for the narrow nail marks and droppings of these birds in the debris.

Beds

Deer feed quickly and then move to a bedding site where they "chew their cud," that is, they regurgitate what they have browsed and rechew it, mixing it with digestive enzymes in their saliva. Feeding in this way exposes them to less danger from large predators since a standing deer in a feeding zone is much more conspicuous than a deer lying down in shadow or brush. These beds are located in varied situations. During windy periods in the winter deer may opt for the still air of bottomlands while in summer such conditions may bring them up on the ridges where the breeze gives them some relief from insects. In settled, windless conditions in winter, beds may be found along the "military crest" of a convex hill, that is, along that contour where the rise of a hill begins to round toward the top. Bedded there, a deer can avoid the pool of cold air in the hollow below. Danger approaching from that direction can easily be detected by simply raising the head and orienting the ears downslope. The animal can then slip off over the convex ridge to its rear without being observed from below.

A favorite winter site for beds is under hemlocks or other evergreen trees that are exposed to the sun during the day. There the snow is shallower and the animal can benefit from both back radiation from the sun-warmed foliage and its own reflected body heat.

The beds themselves, which stand out visibly only in snow, can be surprisingly small. Figure 7.2 H shows one photographed on a ridgetop in shallow snow at Petersham. It measured approximately 2x3 feet with the smooth side outlining the neck, back and rump of the animal and the irregular side representing its drawn-up legs. Scat and urine are often found in the bed along with tracks made as the animal stood up to leave.

Habitat

Any area where there are plants and shrubs of browsable height and secure forested bedding sites within a few hundred yards may harbor white-tailed deer. Beaver ponds provide excellent habitat, as well, affording edge brush and emergent plants. Deer may even wade into the pond in summer to feed on lily pads. Powerlines provide brushy forage and sunny exposures as well as routes of dispersal for deer and many other animals. Shrub growth along the edge of a woodline is often visited. Incuts made into such a woodline by suburban development increase the linear edge, or "ecotone," of a woodland. Responding to the open sky, brush may grow back along this edge providing, along with ornamental shrubs, a strong inducement for twilight visits by deer.

Trackard 8 – Eastern Coyote

The Eastern coyote is a larger version of its more familiar western cousin. In appearance it has some wolf-like characteristics including a thicker muzzle and a heavier, more muscular body. One trait it shares with its western counterpart, however, is intelligence. Coyotes are among the cleverest and most adaptable animals in the woods, and more recently in the suburbs of eastern cities, able to insinuate themselves around human schedules and activities with such discretion that their presence is rarely suspected. A familiarity with their tracks and sign will reveal this presence as much more common than is generally thought.

Eastern coyotes are usually about as big as a small-to-medium-size German shepherd with which they often share similar blond, grizzled coloration. While the shepherd's tail describes a slender inverted arc, however, the coyote's is a large, black-tipped brush that is held down while the animal walks or trots. Also, reddish guard hairs on the coat of many blond coyotes give the pelt a blush that is usually absent in shepherds. Several color morphs exist, however, including gray (the "silver coyote") and even black. These morphs and the large size of the Eastern coyote gave rise to a theory among hunters that this animal was a "coydog," a cross with domestic dogs. The random gestation cycles of domestic dogs, which bear pups in inappropriate seasons, the fact that male dogs do not help in raising pups and, finally, the lack of winter efficiency honed by evolutionary selection make it unlikely that offspring of a coyote-dog union would survive for long in the wild. Following the trail of an Eastern coyote in the snow for any distance shows an animal exquisitely attuned to its environment and to efficient movement through it. Following the profligate trail of a domestic dog, on the other hand, reveals an animal mainly adapted to a can of dog food at the end of the day. This difference alone is likely to guarantee a lack of long term viability for any coyote-domestic dog cross, the instincts and characteristics of which have been adulterated by long association with humans.

Tracks

Coyote tracks show the usual canid characteristics of lateral symmetry and a distribution of toe pads clustered around a central pyramid shaped like a five-point "star." Four toes register in both front and hind prints, the fifth vestigial toe having migrated through evolution so far up the foot that it seldom shows in tracks. Coyotes are digitigrade, walking on their toes and secondary pads. The rest of the foot shows only when the animal was sitting or lying down. The foot is remarkably rigid, having been designed for running, digging and little else. Toe dexterity is not typically a canid virtue, and so coyote tracks in any given condition will vary little from step to step.

Because the front feet support the weight of the animal's head, front prints are slightly larger than hind. This can be seen in the prints shown in Figure 8.1 A and B, recorded in thin snow in Princeton, Massachusetts. Since coyotes dig as part of their living, their nails are worn down more than are those of domestic dogs. The middle two nails register lightly in the print, but the medial and lateral nails, hidden in interdigital fur, usually do not, a feature that is often diagnostic. In walking and trotting patterns on a firm surface, the toes are tightly closed, especially on cold surfaces where such an arrangement conserves heat. The middle two toes describe ovals while the medial and lateral ones often show an angular interior margin typical on many canids including domestic dogs. This angularity, which gives these pad impressions the shape of a large slice of pie, can be seen clearly in several prints on figures 8.1 and 8.2. The front secondary pad is quite large; in prints with toes closed its breadth will always be wider than the middle two toes. The resulting fore and aft asymmetry is the reason one is seldom in doubt at first glance as to which way the animal was headed, however vague the details of the track may be (compare with Trackard 14 – Red Fox). In addition, this pad often shows distinct corners on the medial and lateral ends. The trailing edge of the secondary pad describes a smooth arc except in some direct registrations where a scalloping on this trailing edge may be detected, caused by the spreading out of the round secondary pad of the hind impressed over the front.

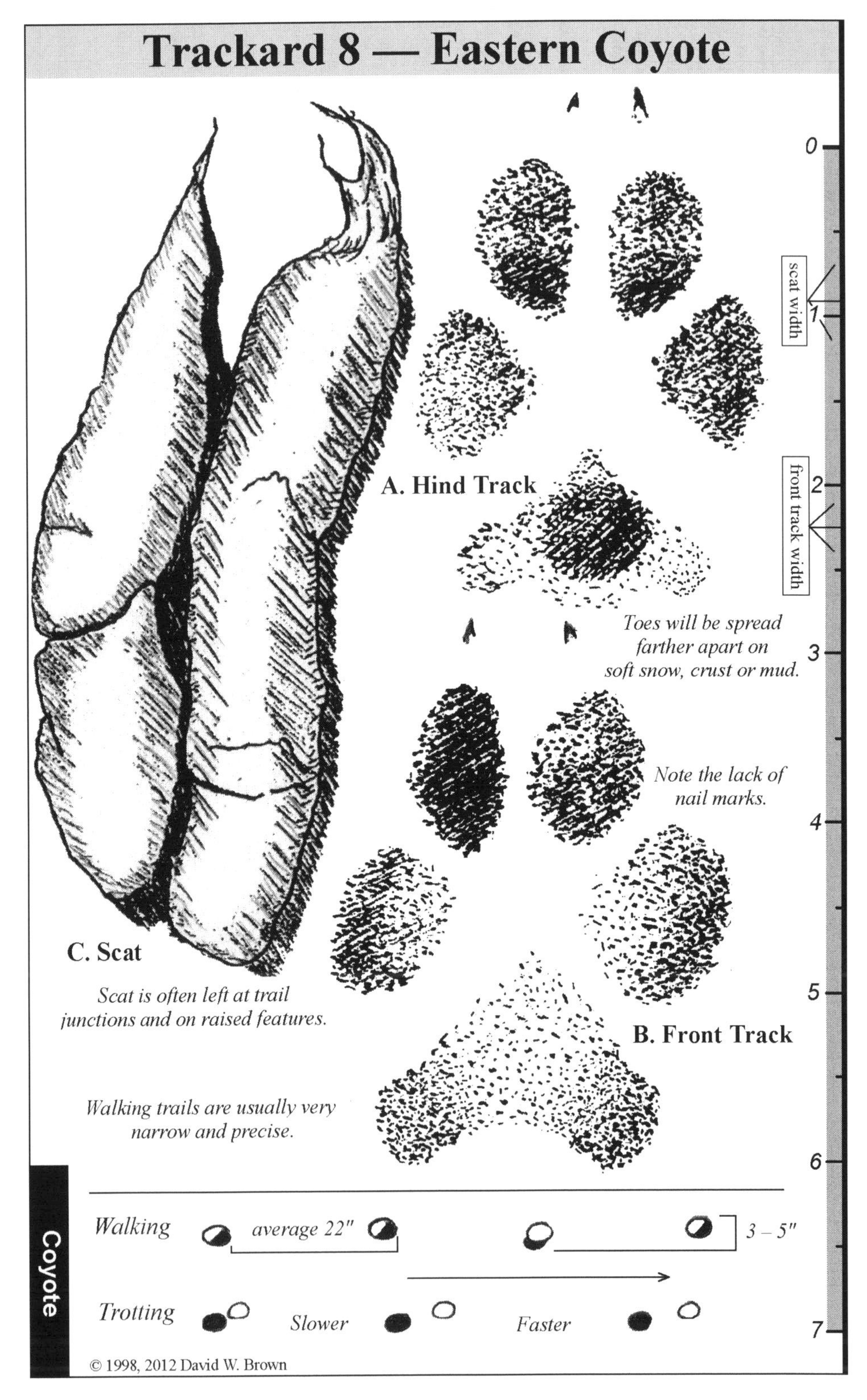

Figure 8.1. Trackard 8 – Eastern Coyote.

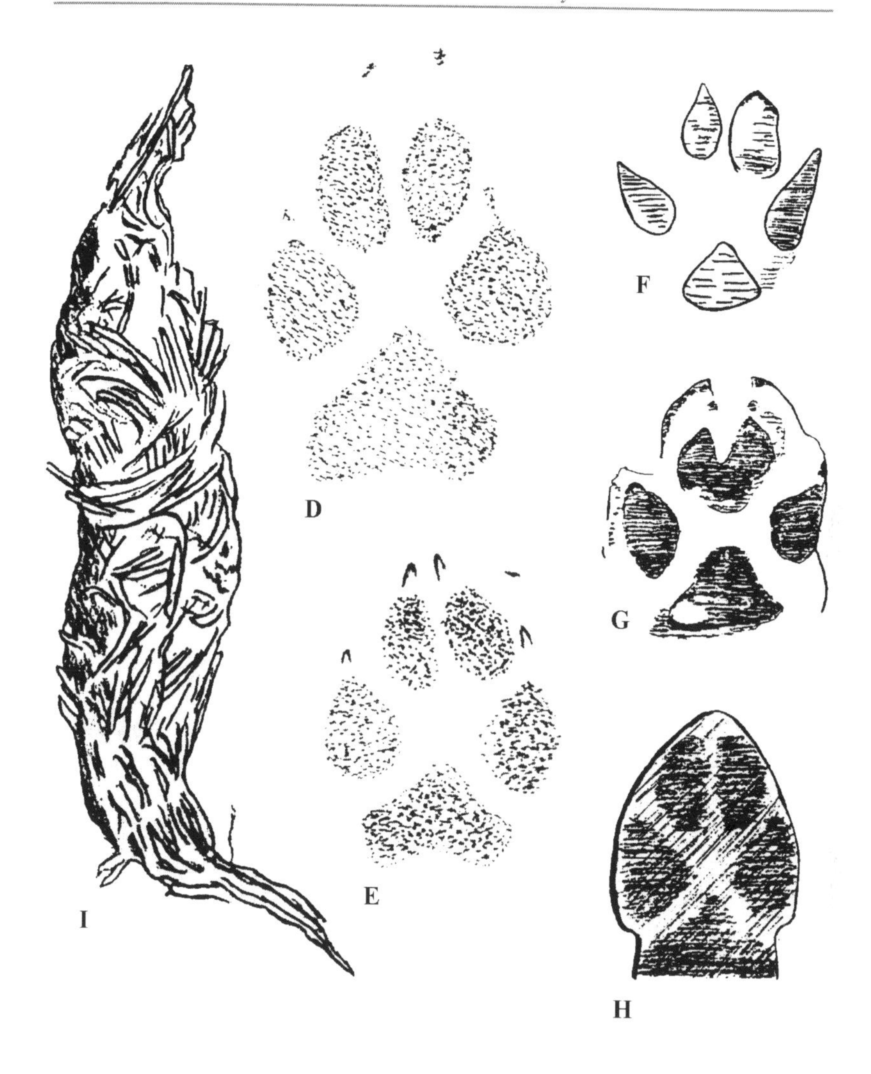

Figure 8.2. Eastern coyote tracks and scat.

D. Front print in snow, about 2/3 size. Princeton, Massachusetts.

E. Front print with a different profile, about 2/5 size. Jackson, New Hampshire.

F. Track in collapsible spring snow. Petersham, Massachusetts.

G. Another direct registration in similar conditions.

H. "Arrowhead" track frozen into snow and slush.

I. A grass scat, about actual size. Jackson, New Hampshire.

When the details of coyote tracks in snow are vague, the closed toes and relatively wide secondary pad often give the impression of an arrowhead pointed in the direction of travel as shown in Figure 8.2 H. This appearance is distinctive to coyote.

The Eastern coyote's hind print is smaller than the front since it bears only the weight of the animal's haunches. The round central area of the hind secondary pad is protuberant and often registers without its lateral and medial wings. Thus this pad usually registers lightly as a vague disc. In deep impressions, however, the wings of this circle may appear faintly, as may a rocker cradling it at its posterior. These effects are clear in Figure 8.1 A.

Trails

An adult eastern coyote walking on a firm, even surface records a step length of about 22 inches with a tight distribution, usually plus or minus only a couple of inches. Registrations are normally direct, that is, the hind impresses perfectly within the outlines of the front. The front foot, carrying the weight of the head, spreads slightly, leaving a larger and deeper impression that dominates the appearance of a direct registration. Occasionally the coyote indirect registers, often laterally; that is, the hind foot impresses slightly to one side or the other, leaving an oblong track. This is not done often, however, and a look down the rest of the trail will show a line of near-perfect direct registrations. The trail width of a coyote, like that of foxes, is quite narrow. In soft, wet thaw snow it may widen to 6 inches, but in most conditions it will average around 4 and will leave an impression of neatness and efficiency similar to that of a red fox.

On hardened surfaces such as a plowed road or firm crust coyotes invariably travel in a "displaced trot," the pattern for which is shown at the extreme bottom of Trackard 8. In this pattern the front prints are lined up on one side of the trail and retarded in each two-print pattern while the hind prints line up on the other side and are advanced. As coyotes often cleverly and boldly use smooth human thoroughfares when humans are not around, this pattern is found often. It is illustrated with increasing speed from left to right. As the coyote trots faster, the pattern spreads out, as does the step-length and inter-pattern distance.

Coyotes can lope and gallop, at least for short distances. These gaits are represented by the same three- and four-print patterns as other animals of similar habit and anatomy like red foxes (see Figure 14.2 in Red Fox). Coyotes are either smarter or less excitable than red foxes, however, and tend to use these energetic gaits more sparingly, preferring to walk or trot most of the time as they move around their range. Although they may stalk a vole on occasion, their usual hunting method is to put in miles hoping to detect the presence of prey with their exquisitely tuned senses.

In the winter coyote packs consisting of the parents, grown juveniles of that summer's litter, and perhaps a female helper from a previous litter who has assisted with their rearing, will often travel and hunt together, their trails sometimes forming troughs in the snow. At this season their trails shadow the movements of deer, even using the packed trails of these animals to facilitate getting round to hunt small prey while they wait patiently for starvation to take its toll on their benefactors. In January and February, adult coyotes pair up to mate. In deep, soft snow their trails often coincide perfectly, with the tracks of the female superimposed over those of her male consort, giving the impression of the passage of a single animal. Following such a trail will eventually show a place where one or the other will give up the truth by diverging to investigate something off to one side, to sniff a fallen log, for instance, before coming back to the trail of its mate. The pair may also deposit side-by-side scat and urine, the female's urine sometimes showing estrus blood in the process.

Track and Trail Comparisons

Red fox: Although they share the typical canid profile with red fox, Eastern coyote tracks are larger, the pads more distinct, and the secondary pad wider than the two central toe pads, at least in usual situations where the toes are closed together. This is so much so that the resulting front print, which usually dominates direct registrations, is asymmetrical fore and aft. Closed red fox prints and tracks, on the other hand, are symmetrical fore and aft, so that at first glance one is often unsure of the animal's direction.

Red fox walking trails usually show a step length of around 15 inches with a very tight distribution of only an inch or two compared to the 22–23-inch step of an adult coyote. However, young coyotes in December snow before they attain full growth may show print sizes and step lengths near red fox size. But at this season, coyote juveniles travel together, unlike red foxes, which by then are on their own, and so the tracks of more than one coyote may be present. Furthermore, coyotes do not show hairy pads at any season. Instead the naked pads tend to show distinct outlines, even in fresh, dry powder while red fox pads in such conditions are diffuse due to a covering of fur. Finally, young coyotes at the beginning of their first winter often show inefficient, puppy-like behavior in their trails, quite unlike the sobriety of any adult red fox with which they might be confused.

Deer: In melted-out tracks such as the "arrowhead" in Figure 8.2 H, a coyote can appear remarkably like a deer. Look for a central mound in such a coyote track where a melted deer print will show a vague ridge running its length. A careful look at the coyote's print may show a slight concavity along its forward quarter representing the space between a middle toe and an outside one. A track of a white-tailed deer will show an even curve in this area. Mule deer, on the other hand, may also show a flattening or even a slight concavity along the side, especially in the hind print.

Bobcat: Bobcat tracks are smaller than Eastern coyote tracks, rounder, usually without nail marks and with a secondary pad covering the same area as three of the toe pads to about two for a coyote or other canid. The flexible toes of a cat often causes the prints to vary their shape from step to step, especially on uneven surfaces, and frequently show lateral asymmetry where each side of a coyote's print is nearly a mirror image of the other. In addition, the medial and lateral pads of a bobcat have the same oval shape as the middle two pads. However, some care must be taken with bobcat direct registrations: where the more symmetrical hind print dominates the appearance of a track, a bobcat's can look quite canid, that is, somewhat elongated, with a central mound and a secondary pad smaller than the front. In such tracks a large bobcat easily may be mistaken for a coyote.

Domestic dog: Most domestic dog prints show splayed pads rather than the tight grouping usual for coyote and, unless they have been clipped, prominent nails on all four toes. It should be noted, however, that coyotes can show splayed toes in certain conditions; on soft, wet spring snow where the surface threatens to collapse underfoot or on soft mud, the coyote will spread its toes for support. The resulting tracks shown in Figure 8.2 F and G can look very much like that of a dog or small wolf. Note, however, that nail marks are not distinct and that the registrations are still perfect; both illustrations are direct registrations of hind into front. Dogs not only show distinct nail marks, but also the registrations tend to be indirect or even double. If more than one animal is involved, as is often the case with wandering or feral dogs as well as coyotes, the print sizes and shapes are more likely to be uniform from animal to animal with coyotes than with dogs, which are likely to be of different breeds.

Scat

Figure 8.1 C, from a photograph taken in November in Princeton, shows classic Eastern coyote scat: tubular, over ¾ inch in diameter, gray and with a "tail" at one end of each segment. This is carnivore scat, the tail being the excreted hair of the prey and the gray perhaps due to calcium from the victim's bones. However, deposits may also be brown or yellowish and often show large bone chips on their surface. With age, coyote scats often acquire a characteristic white patina, presumably due to alkalinity that itself is the result of the scats' calcium content.

Coyote scat is frequently left on human trails and lightly used roads as a signal of the animal's passage. Most frequently it is found at intersections or in a raised portion of the trail, sometimes in several deposits with different ages suggesting communication among a number of animals. At these intersections I often find it placed slightly off-center toward the lesser used trail, possibly in an effort to avoid its disturbance by foot traffic.

Close to civilization, coyote scat can contain a wide variety of food items derived from human trash. During late summer it is often composed of the seeds of a single variety of whatever berry is fruiting at the moment. Coyotes like grapes and fallen apples; the resulting scat will show appropriate color and seeds. Scat may also be composed of grass, consumed to stimulate the bowel. If the scat is fresh and the grass was green when it was eaten, the scat can have the appearance of a Cuban cigar. Figure 8.2 I, on the other hand, shows a scat from late fall when the grass was yellow and dried. Pine needles may also be consumed off the forest floor or off a downed limb, presumably for the same reason. In April when coyotes are shedding their heavy winter undercoat, they may bite off the clumps and swallow them. The resulting scat, composed of the animal's own hair, may have a yellowish appearance matching its coat. At mid-winter when coyotes are scavenging old carcasses, the resulting scat also acquires a dry, blond appearance; the carcass may have so little flesh on it that its remnants pass through the coyote's gut relatively unaffected by bacteria and the resulting scat lacks the dark mucous of scat from more nutritious feeding.

Scat Comparisons

Red fox: Red foxes and coyotes are both territorial carnivores that often leave scat on trails. However, deposits of coyote scat are more copious than those of the small-stomached fox and larger in diameter. Red fox scats in the Northeast are normally less than ¾ inch in diameter with ½ inch usual, compared to ¾ inch or more for coyotes. However, sometimes foxes get constipated; the resulting deposit, while wider than normal, is usually in a series of short, discreet lumps and is never copious. Juvenile coyotes may leave deposits the size of which overlaps with that of adult foxes. However, coyote fur-scat is more uniformly tubular while fox scat often has constrictions and connected lumps. Finally, fox scat will not show large bone chips on its surface while coyote scats often do, especially if from late stages in the reduction of a carcass.

Domestic dog: Dogs commonly leave deposits on the trails that they walk with their masters. The revolting odor of domestic dog droppings, however, is not present in the scat of wild canids, perhaps because of a more refined digestive system in the wild animals that leaves less in their scats of interest to bacteria. The consistency of domestic dog scat is uniform, showing the remains of grain, which is the main constituent in commercial dog food.

Black bear: Young black bears feeding on fallen apples may leave scat that is very similar in size and appearance to that of Eastern coyotes frequenting the same food source. Coyotes usually approach and leave orchards over farm roads and trails while bears are more likely to come through the woods. Thus the location of a scat may help in determining which was the depositor. Coyotes also may urinate and scrape near the scat deposit. Finally, tubular segments of coyote scat (as opposed to teardrop-shaped segments) tend to have a uniform diameter over most of their length while bear scats often show a characteristic taper on one or more of the segments (see Figure 1.2 H in Black Bear).

Sign Comparisons

Coyotes scent-mark with their urine much as do foxes and domestic dogs. However, the smell of a coyote's is less acrid and intense than that of red fox. It has the odor of mild skunk mixed with burnt hair. Also, unlike the scent of red fox, it does not travel far and, when found, is often associated with scat or scratched up ground. Domestic dog urine is normally odorless and paler than coyote urine, at least in winter when coyotes tend to leave more concentrated deposits.

Coyotes are social animals, with families often getting together for a convocation at trail intersections, on beaver ponds and at other open sites. At these locations they may group-howl and deposit much scat and urine. Like domestic dogs, they often scratch violently with their back feet beside a urine or scat deposit. In the winter these same family packs frequently hunt together although without the unified direction and close coordination of a wolf pack. Their association at this season seems partly for efficiency in moving through deep snow and partly as a feeding convenience. Eastern coyotes are dependent to a certain extent on winter deer mortality for survival. When a deer goes down from starvation, disease, exhaustion or old age, the coyote pack, which may have helped in the final stage of its demise, arrives at the carcass. If the group is a family that has already sorted out its dominance structure and feeding order, the danger of violent confrontation is reduced and with it the possibility of injury, which in the winter would be a death sentence. The carcass is reduced in an orderly and efficient manner, with nothing left to waste. As the carcass falls apart, individual sub-dominant animals may carry off parts and cache them in secret where they will not be forced to submit to the family's feeding order.

A deer that has been killed by feral dogs, on the other hand, looks a mess. Dogs have lost the instinct for orderly and efficient feeding. An inventory of the tracks of a feral dog pack in the vicinity of a carcass will usually show a mix of sizes and shapes due to the variety of breeds of which a pack typically consists.

One place where coyote pack organization does contribute to hunting success is in situations where the group tries to drive a deer out onto the slippery ice of a pond or lake. By working as a team the coyotes can cut off possible escape down the shoreline. Once even a healthy deer is out on the ice, it is helpless. It cannot rear up on its hind legs and strike out at its attackers. Instead it eventually spraddles and falls, becoming easy prey for the pack.

Intelligent and opportunistic, coyotes have learned to take advantage of the huge increase in Canada goose populations in the East.

Lying hidden near the water's edge until a goose comes within range, the coyote plunges in and seizes the bird by the neck, dragging it out onto the shore where it can feed. As a bird this size has few natural predators, the remains of a Canada goose on a bank can generally be taken as a sign of coyote predation.

Dens

Coyotes den from March through July and then only to bear and rear pups. However, at other seasons young animals may return every now and then to the remembered security and familiarity of their natal den area to feed or to take shelter during a storm. The den is often a renovated woodchuck tunnel, frequently on a south-facing slope for warmth and drainage and in a place where human or domestic dog intrusion is unlikely. The entrance is enlarged to 1½ to 2 feet in diameter, a feature that immediately sets it apart from other animals, at least in the East. On the debris pile at its mouth will be found scats and the remains of prey delivered to the pups at the den by the parents, possibly assisted by a helper juvenile from the previous litter. Coyotes often have alternate dens to which they may move the pups if the natal den is disturbed or becomes infested with vermin. A move may also be made to a cooler site as spring turns to summer.

Habitat

Eastern coyotes will live in any habitat that affords them a food source and a concealed denning site for raising pups. In snow country they are more dependent on deer for late-winter food although they will hunt voles under the snow as well as other prey while waiting for a weakening of individuals in the deer herd. At other seasons they are more omnivorous, feeding on almost anything edible within their home range. One particularly good habitat is the center strip of interstate highways where rodents are plentiful and humans and their dogs are totally absent. In a winter of deep, unconsolidated snow where pursuit of deer and other prey is difficult, they may turn to raiding livestock, especially if a herd of sheep, for instance, is left unprotected by man or dog near a woodline that affords concealed approach and departure. Efforts at controlling coyote depredation have been various and strenuous, sometimes, as in the case of poisoning, doing a great deal of ecological damage as the toxins spread through the food web. The end result with respect to the coyotes, however, has never been success, at least not in the long run. Coyotes have the ability to adjust litter-size to the availability of food, quickly replacing any of their number lost to human predation.

It should also be noted that many other wild animals depend on successful hunting by coyotes and other large predators for winter survival. The remains of a winter deer kill, for instance, are scavenged by many birds and mammals. Hawks, foxes and even bobcats feed off the remnants directly or prey upon animals that concentrate at the site. Even snowshoe hares are reported to gnaw frozen meat. In the winter all animals are being tested by starvation, and a coyote kill may be all that stands in the way of death.

To the traditional coyote attributes of cleverness and opportunism are added, in the Eastern version, size, strength and athleticism. Several centuries of persecution have succeeded only in developing, particularly in this race of the animal, a super-coyote with a high level of strength, intelligence and resourcefulness. As long as there is suitable habitat for this adaptable animal, the coyote will be with us, a ghostly presence howling on a moonlit night.

Trackard 9 – Porcupine

Porcupines are rodents that have evolved a unique defense; some strands of their fur have been modified into quills covered with overlapping scales that point backwards from the tip like barbs on a fish-hook. Anything that molests the animal runs the risk of getting some of these imbedded in its hide as punishment. Porcupines cannot throw their quills, but the lightning-quick strike of the animal's tail can make it seem so. Not only does Nature abhor a perfect predator, which would soon eat itself out of its food supply, but it also abhors a perfect defense for the same reason. As a result, the porcupine's defense has been provided with a chink; most quills are on its rear end, which it must keep directed toward any attacker. This defect makes it vulnerable to several predators but especially to the fisher, which attacks the animal's face until blood blinds and disorients it to the direction of the attack, rendering it helpless. A wise porcupine in fisher country hides its head in a crevice of some sort where this fierce predator cannot get at it.

Porcupines can be surprisingly vocal. The presence of an interloper or the urgency of mating may provoke loud moaning or high-pitched "singing" seemingly out of character for this stolid creature.

Tracks

Porcupines are plantigrade walkers, that is, the heel of the foot registers with each step. As shown in Figure 9.1 A and B, based on snow prints photographed in Petersham, Massachusetts, the secondary and tertiary pads are fused into one large area akin to the sole of the human foot. This fused pad is quite thick and covered with rubbery protrusions for traction. Porcupines show five nails in the larger hind print but only four on the smaller front, contrary to information in some tracking guides. The nails, used both for digging in the litter of the forest floor and for climbing, are quite long especially on the front foot. In vague prints their collective points often register as an

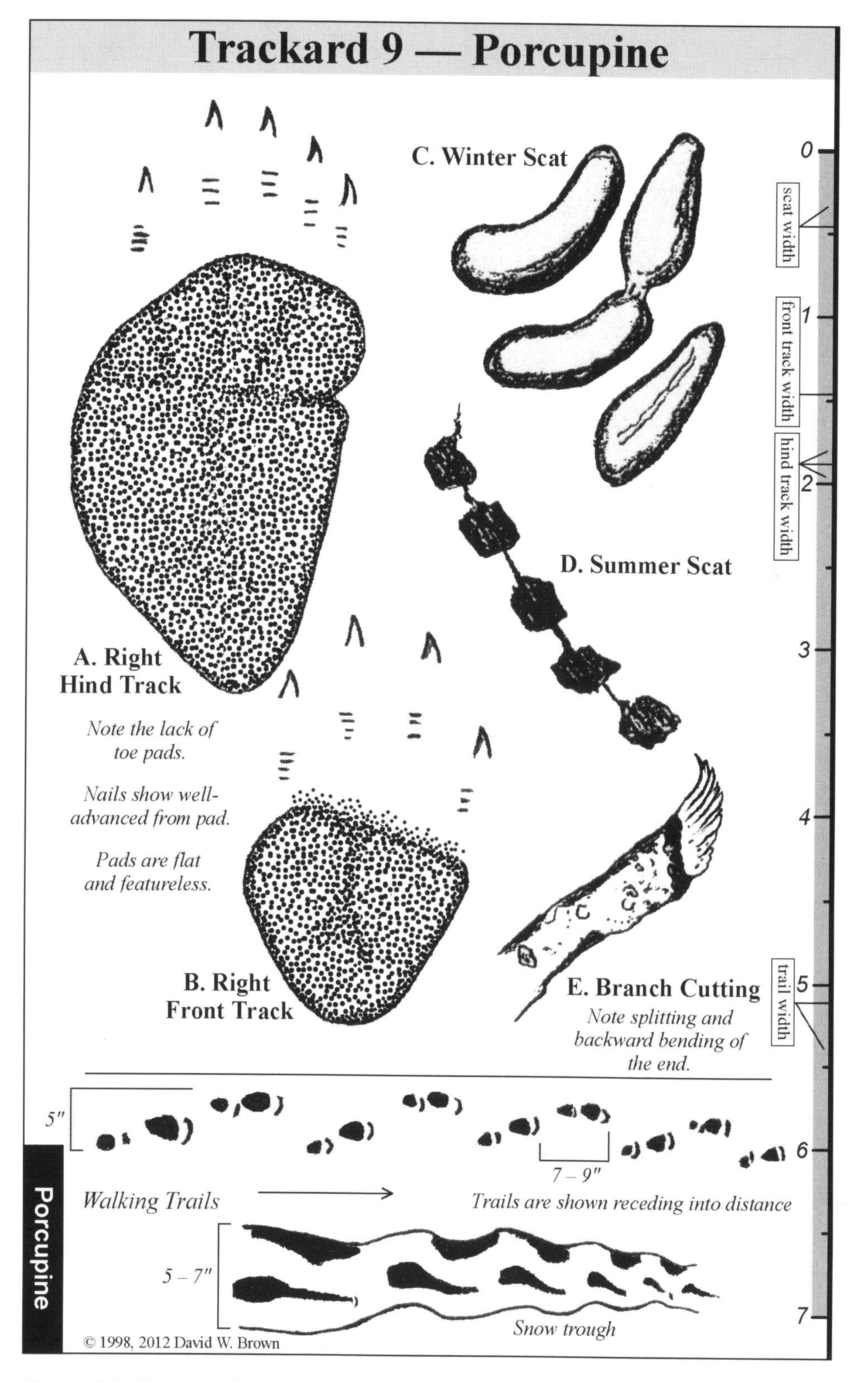

Figure 9.1. Trackard 9 – Porcupine.

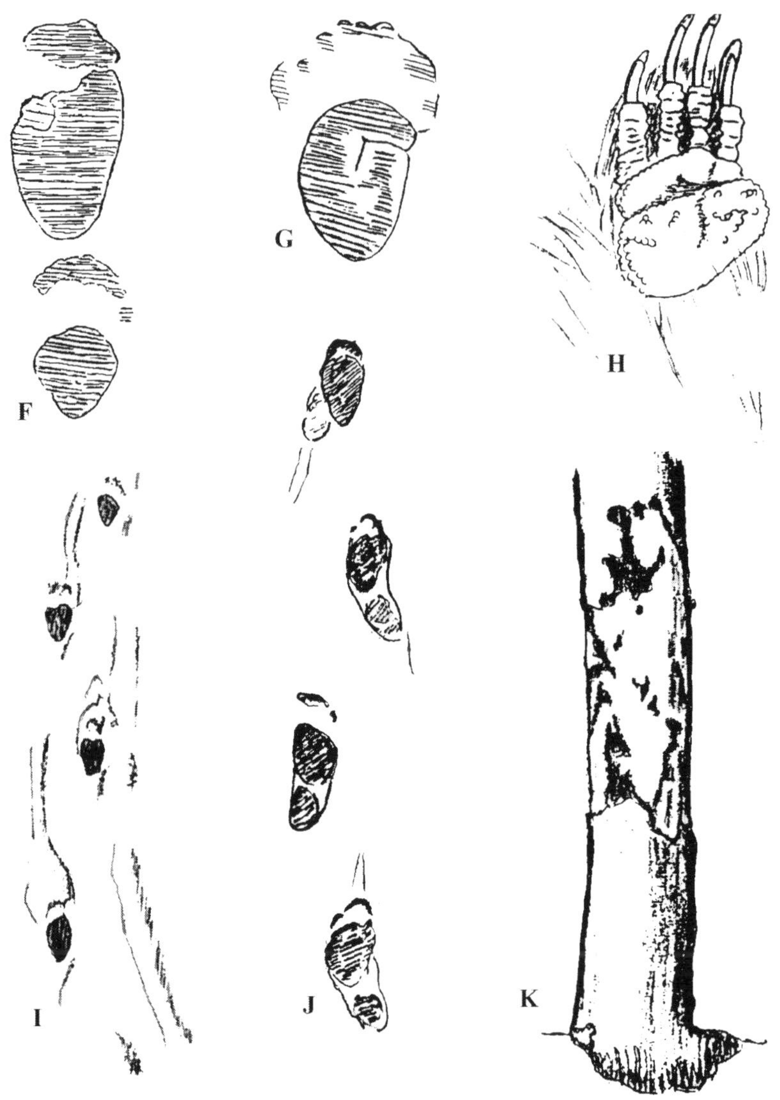

Figure 9.2. Porcupine tracks and sign, all reduced.

F. Shallow hind (top) and front prints in thin snow over ice. New Salem, Massachusetts.

G. Another shallow hind print, showing wide nail arc.

H. Front foot from a fisher kill site. New Salem, Massachusetts.

I. Direct register walking trail in snow with nail drag. Concord, Massachusetts.

J. "Reversed comma" overstep walk with front prints behind and outside hind. Petersham, Massachusetts.

K. Debarking of young white pine. New Salem, Massachusetts.

arc quite far forward of the sole of the print, so much so that they can become disassociated to the eye and easily missed. This effect is clearest in the lower (front) print shown in Figure 9.2 F, recorded on a plowed road in New Salem, Massachusetts. It is reinforced by the fact that intervening toe impressions are withdrawn from most tracks. As seen in Figure 9.2 H, from a photograph of a porcupine's front foot, the curve of the nails and the thickness of the footpad combine to hold the toes above the plane of most prints. In fact, in vague tracks, such as those found in loose sand for instance, the combination of a well-advanced crescent of nail tip-marks followed by a wide, vacant space and then a flat, featureless oval impression is diagnostic for this species.

The hind print is about twice as large as the front with its sole shaped like the letter D. Because of the rubbery thickness of the sole, however, this pad may be distorted into an oval in some prints such as in Figure 9.2 F and G. Whatever shape it takes, the print made by the sole is very flat with well-delineated edges. The small front print often appears as just a round impression near or partly under the heel area of the hind and with its nail marks obscured by the larger print.

The texture of a porcupine's footpads is knubbly, for traction on the bark of trees or even the rock of cliff faces. In very clear prints in mud, this leaves a characteristic pebbled appearance.

Trails

A typical walking pattern of a porcupine is illustrated in the trail at the bottom of Trackard 9 (Figure 9.1), shown receding into the distance. In this case the overstep was considerable and so the front prints were discreet single registrations. Note that the straddle of the front prints is wider than the hind; a porcupine climbs a tree by embracing it with its front legs and pushing itself upward with the hind, an action requiring a broader chest than pelvis. Thus, in trails where the overstep is only slight, the front print appears just behind and to the outside of each hind print. The result is a series of distinctive comma-shaped double registrations, each one the reverse of the previous as in Figure 9.2 J, recorded in slush ice under thaw snow at Petersham, Massachusetts.

In deep soft snow, the waddling of a walking porcupine translates into a distinctive wavy trough about 6 inches across such as the one at the extreme bottom of Trackard 9, from Concord, Massachusetts. As each foot lifts out and moves ahead, the long nails register as a drag mark forward of the print. Another example with clearer tracks is shown in Figure 9.2 I, also from Concord. On firm snow these nail marks may also register, but as thin striations forward of the track. Troughs may be used repeatedly until they become packed-down trails between den and conifer.

Besides walking, a porcupine, agitated by the presence of a rival or the proximity of a threat, can also lope, but such trails are uncommon. Most of the time porcupines walk, having less need than other rodents to move faster, given the effectiveness of their defense.

In warmer months porcupines often follow the same routes between den and feeding tree so repeatedly that a vague trail wears into the duff of the forest floor. It will be about 5 inches wide and will go around small obstacles. A deer trail of similar width will go over logs and boulders and, unlike the trough created by hares, the porcupine's will usually end at a tree. Finally, on a firm dirt road where individual prints do not register, one can sometimes make out crisscross scratches left by the tail quills of the waddling animal.

Scat

Figure 9.1 C shows typical winter scat of porcupines. The discreet pieces are the sort left where the animal has been gnawing the bark of pine or other resinous trees. Often such scat shows a seam on its concave surface, as in the bottom piece in the illustration, along with the distinctive odor of pinesap. The connected pieces shown in this cluster of pellets are from the edge of a field where the animals were feeding partly on grass just greening after April thaw. Where porcupine scat accumulates, a search will often reveal a quill or two lying among the pieces.

Porcupines tolerate a diet of conifer parts in the winter because they have no choice. As soon as spring arrives, however, they switch to the buds and flowers of hardwoods wherever they are available. The black or brown string of "train scats" in Figure 9.1 D is the sort sometimes found where the animal has been feeding on succulent forage. The illustrated example, located under a white oak in New Salem, Massachusetts, shows a series of lumps connected by a membrane of mucous and fibers.

Porcupines feed for long periods in the same tree, alternately eating and dozing. Where fresh scat is found, a look overhead may reveal a dark mass along a limb or in a crook of the trunk and lead to the discovery of the animal itself.

Dens

Porcupines typically den in the crevices of rocky ravines, in hollow trees or in culverts. These dens are easy to locate by sight or smell. Outside the entrance to each will be found pellets of winter scat which, along with pungent urinations, mark their tenancy with an unmistakable piney odor. Sometimes these scat piles achieve huge proportions of what may appear to be millions of individual droppings. Porcupines will use any convenient hollow for a den such as an old beaver bank burrow that has partly collapsed providing access from above. A bank lodge of the same animal, which has been abandoned in a drought due to the receding shoreline, may also be used, or even a fallen stovepipe in an old lumber camp, anything that provides a porcupine with protection while it rests.

Sign

Stunted hemlocks show the reason fishers were artificially reintroduced in northern New England by foresters in the 1940s. Porcupines feed on fresh twig tips at the top of trees, frustrating vertical growth and resulting in a tree with a short, squat shape. Also patchy debarking by these animals exposes a tree to infection. Figure 9.2 K, from New Salem, shows this gnawing on the smooth bark of a young pine. Following up the trunk with one's eye may reveal a series of such patches, one at each spot where the animal could find a comfortable place to sit while it fed.

Sprays of hemlock branches may be found littering the snow under stands of this tree species, with packed trails leading back to the porcupine's den. The cuts porcupines make on branches are distinctive. The animal crawls as far out on the limb as its weight will allow. Then it stretches out a front paw, hooking its claws over the branch, and begins making diagonal cuts with its central incisors. The weakening of the branch is detected by the leading paw which then draws it back to the porcupine's body for control. After the last cuts are made, the animal feeds on the end growth and lets the spray fall. The stressing of the branch during the last cuts results in characteristic splitting or "bend-back" of the diagonally cut end shown in Figure 9.1 E, a branch found under a red oak in New Salem.

Porcupines like salt and have a deserved reputation for demolishing the wooden handles of axes and other tools used by humans for sweaty labor. Even the seats of outhouses, which regularly contact human skin, will be gnawed. They also like the taste of paint and will neatly "debark" a trail signpost for this substance.

Sign Comparisons

Beaver: When beavers cut branches, they are standing on firm ground, with no need to control the fall of the branch. As a result the end usually does not show stress splitting, the cuts instead being neatly chiseled.

Squirrel: In the late summer and fall, squirrels also crawl out on branches and cut off the end twigs to harvest buds, acorns or cones. Because neither porcupine nor squirrel does more work for its food than it has to, in this case preferring to cut as thin a branch as possible, they both go as far out on the narrowing branch as they dare before cutting. Thus the ends of fallen twigs nipped by the lighter squirrel will be narrower than those of the bulkier porcupine, which must cut closer to the trunk.

Deer and **bear:** In early winter, before deep snow has accumulated, porcupines often tear up large patches of oak leaves looking for acorns missed by or buried by squirrels. Such work can look like that of deer or bear, both of which do the same. Lacking nimble front toes with which to manipulate the acorn, porcupines just crunch it and extricate as much meat as they can, leaving what they can't inside the crushed and discarded shells. Bears crunch and swallow the whole acorn leaving nothing behind but their scat. Acorns crunched by deer, on the other hand, can look very much like porcupine work. Careful examination may be necessary in the softened areas of torn up duff to find tracks, scat or quills.

When porcupines climb spindly saplings for leafage or spring flowers, they sometimes overtop the tree and ride it to the ground, breaking its top. This work can be distinguished from bear or moose sign in several ways. Bears break branches and small trees for fruit or berries, not leaves or flowers; look carefully at what has been eaten. Moose do browse the leafage but the twig ends will show the usual ungulate tears rather than the neater diagonal cuts of rodents. Be careful, however; once the treetop has been brought to the ground by the porcupine, its twigs may be browsed subsequently by other animals.

Two incisor marks close together on bark always mean a rodent or lagomorph and quills loosened by the fall will narrow the choices to one.

Habitat

In winter, porcupines keep to groves of conifers for most of their feeding although they may debark associated hardwoods such as beeches as well. In early spring they are attracted to grassy areas that have remained green under the snow; at dusk one may find several of these normally solitary animals feeding near one another in a field. They may also climb aspens and other trees for their flowers. At such times porcupines may be found near the top of adjacent and impossibly spindly saplings, singing at each other nervously. In the summer they prefer to forage in hardwoods such as oaks, where they seem to feed on the least tannic parts. From late summer to early winter acorns are a staple. In a poor acorn year, I have found porcupines eating the green leaves of red oaks as a substitute, leaving behind only a skeleton of leaf veins. The leaves of white oaks are avoided for some reason although their acorns are preferred. Perhaps the tannin in white oaks, largely absent in their acorns, is concentrated in their leaves instead. So under white oaks you may find sprays with intact leaves but without acorns while under adjacent red oaks you may find caterpillar-like leaf feeding.

Trackard 10 – Raccoon

Raccoons are intelligent and resourceful omnivores, able to live on anything edible. Their facial mask, gray fur, humped spine and ringed tail give them a distinct appearance unlike any other animal in the woods. With feet more like hands, they are adept at searching mud for crayfish, mussels and other prey hiding there. Although mainly nocturnal, their hunting forays can last well into the day in woods where few humans and their dogs intrude.

Tracks

Raccoons are the great foolers. Their toes may be spread or closed as well as extended or drawn in. Furthermore, if the animal is up on its toes, the distinctive heel area of the hind print may not register. Thus tracks can have a wide variety of appearances and look like those of a number of other mammals as illustrated by the profiles on Trackard 10 and in this guide (figures 10.1 and 10.2).

Raccoons have five toes on both the front and hind feet; all normally register in prints. When the toes are extended and spread for support as in Figure 10.1 A, recorded in slippery mud along a streambank in Lincoln, Massachusetts, they have a blunt, club-like appearance with parallel sides and perhaps a barely noticeable widening at the tips. This appearance is a signature for raccoon prints in these conditions. However, when the toes are drawn back and the phalangials de-emphasized or withdrawn entirely, the toe pads appear as ovals much as those of many other mammals. Figure 10.1 B, recorded in dry mud at Brownfield, Maine, shows this sort of toe impressions on prints in which the tertiary (heel) area also failed to register. Taken together, then, Figure 10.1 A and B can be thought of as representing the extremes of variation for raccoon prints.

When tertiary pads appear in both prints, the front print will appear smaller than the hind; when they do not register, the two prints may appear nearly the same size as in Figure 10.1 B, or the front,

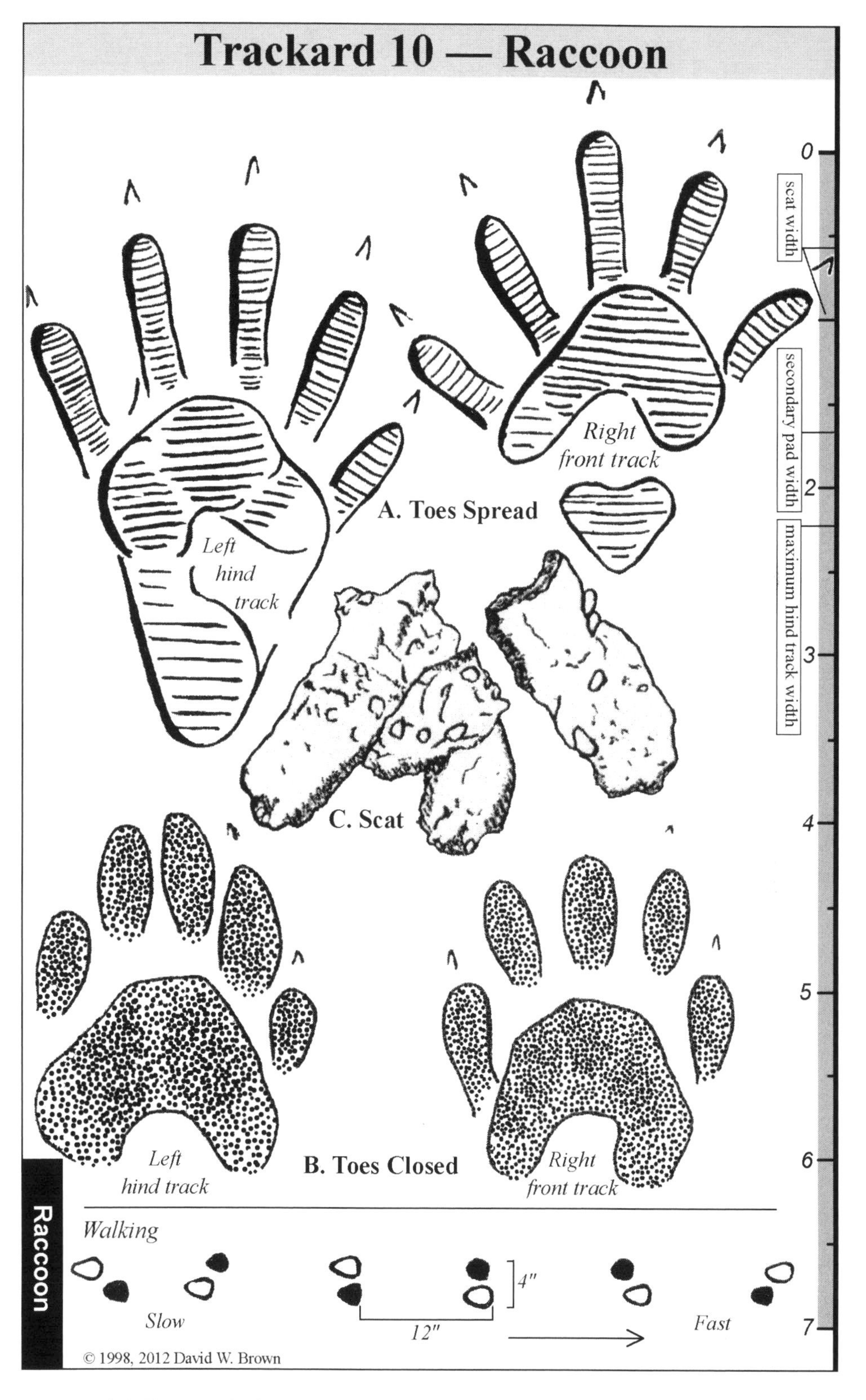

Figure 10.1. Trackard 10 – Raccoon.

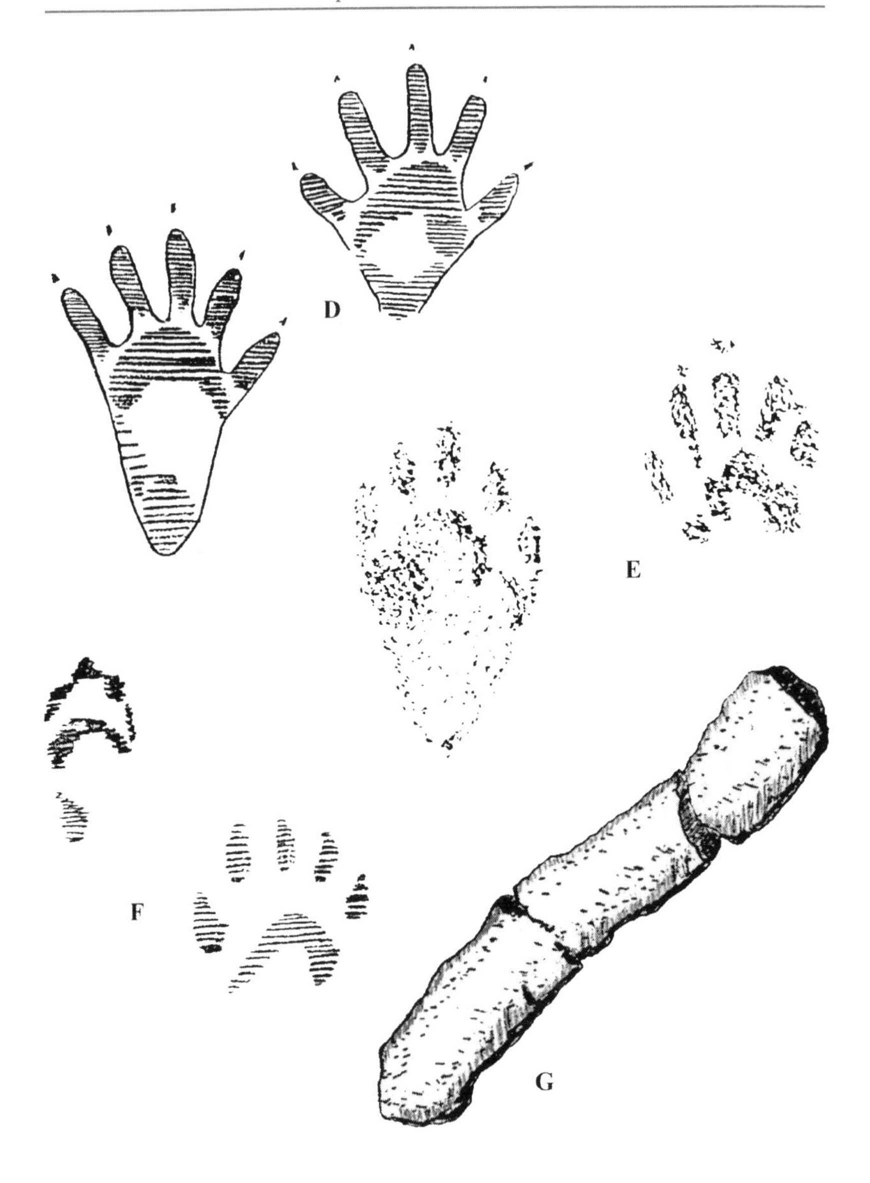

Figure 10.2. Raccoon tracks and scat.

D. Spread prints in snow, about 2/5 size. Saugus, Massachusetts.

E. Contracted prints in thin snow over ice. Lincoln, Massachusetts.

F. Contracted prints showing "fisher" profile of front print on right. Petersham, Massachusetts.

G. Scat drawn from field notes, actual size. Petersham, Massachusetts.

bearing the weight of the animal's head, may be larger as in Figure 10.2 F. However, in distinguishing front from hind, the toes on the front print are more symmetrically arranged, with the medial toe only slightly retarded. The heel area shown in the front prints in figures 10.1 A and 10.2 D registers only occasionally.

Hind prints show a more asymmetrical arrangement of the toes, with the medial toe strongly retarded toward the rear of the print. This is especially clear in Figure 10.2 E from thin snow over ice in Lincoln, Massachusetts. The heel area normally registers in the hind print, making it appear much larger than the front.

In both front and hind prints the secondary area provides another signature impression. It is fused into one smooth pad with a sharply defined leading edge forming an arc behind the toes. This secondary pad is de-emphasized on the medial side, that is, toward the centerline of the animal's body, giving the pad the shape of an old-fashioned laboratory retort. This is particularly clear in Figure 10.2 D. Since raccoons can spread their toes so widely, track width is not as helpful as the more constant secondary pad width of about 3½ centimeters, or a little less than 1½ inches, for an adult.

Trails

When a raccoon is walking, its flexible, humped spine allows it to bring its hind foot all the way forward to the vicinity of the diagonally opposite front print, creating patterns of two prints as shown in the illustration at the bottom of Trackard 10. Such a trail displays a couple of signatures. First, the larger hind print alternates sides with each step. Secondly, in both a slow and a fast walk the front and hind prints are arranged on alternating slants as shown on the extreme left and right of the illustrated trail. Even in a medium speed walk, as in the center of the trail, where the prints are even, the larger hind print still alternates sides with each step.

Raccoons occasionally direct register walk like other wild mammals, but this is usually in deep, soft snow where double packing represents an efficiency. Following such a trail to firmer snow will show a return to the 2X patterns described. Raccoons can also lope, in both the rotary and transverse patterns, as an escape gait or when crossing a danger area. Since these are patterns of single registrations, the differences between front and hind prints noted above should distinguish them from other animals.

Track and Trail Comparisons

Fisher: When a raccoon is traveling on a cold, firm surface, it may contract its toes as in Figure 10.2 F, photographed on a frozen pond in Petersham, Massachusetts, which was covered by a thin layer of firm snow. The result can be very similar to the track of a female fisher. This is more true of the front print, shown on the lower right of the pattern, than the hind, which generally has a different aspect in these conditions. If only the front print is available, however, close attention to the shape and width of the secondary pad may help to tell them apart. Otherwise the alternation of slants and/or print sizes described above will be the distinguishing features.

Muskrat: In confusing prints where the distinctive outer swim flap profile doesn't register, muskrats show toe prints that tend to taper evenly to a point at the ends. If the swim flap does register, identification by the "echo" outline is easy. One can also look for the muskrat's much smaller front prints, if they have not been covered by the hind, as well as for tail marks, neither of which will show in raccoon trails. It is also worth remembering that when raccoons are hunting along water, their trails tend to parallel the shore while muskrat trails tend, more often than not, to run perpendicular to the shore between the safety of water and forage up on the bank.

Woodchuck: Woodchucks have feet about the same size as raccoons. However, they show a typical rodent profile of four toes on the front foot to the raccoon's five. As registrations of all toes are fairly reliable in both species, this should usually tell the difference.

Scat

Since raccoons are omnivores, the appearance of their scat will vary widely with diet. Everything from berry pits to corn skins to trash may be found in it. In one deposit fractionated in Melrose, Massachusetts, I found cardboard, cigarette cellophane, tinfoil and broken glass! The usual appearance is shown in figures 10.1 C and 10.2 G, that is, a segmented cylinder with blunt or squared-off ends. The typical width is about ¾ inch but may be as large as 1 inch.

Location is often a clue to identifying raccoon scat. It is frequently left on a prominence of some sort, such as the hump formed by the roots of a tree encountered along its trail. As raccoons spend a lot of time along the shore where they probe the shallows and mud for prey, the prominent root hump of a tree growing out of the bank may be used repeatedly, forming a scat station for the animal. In any such location one may find a number of scats of different ages and consistencies depending on what the omnivorous raccoon had been eating. The variation in color and texture of similarly shaped deposits in one location, often the base of a tree, is itself a signature for this omnivore.

Sometimes raccoons leave scat at the base of a tree they have climbed to den during the day. The tree may have a cavity in its upper reaches or a broken-off top or any level place where they can doze in safety. If the den is used repeatedly, several scats may accumulate at its base.

Scat Comparisons

In late summer when the berry crop is ripe, raccoon scat may be just a shapeless mass of pits. Although a scat left at an intersection points to coyote as does scratching nearby, there is little to distinguish such scat from that of other mammals feeding on fruit.

Otter: Otters also leave scat deposits on humps near water and live on many of the same prey items as raccoons, such as crayfish and mussels. Large urine burns on moss, pulverized duff from belly scratching, mounded vegetation under scat or a vague trail leading to the water are all indicators of otter rather than raccoon.

Health warning: The dried scat of raccoons, even more than that of other animals, should be regarded with caution. Fractionating dry specimens may release invisible spores of heart roundworm into the air that, if inhaled, can cause serious illness. Also, raccoons sometimes carry the rabies virus. Run away from any wild animal that does not run away from you!

Dens

Besides using dens for birthing and raising young, raccoons often den during the day, especially in areas frequented by humans and dogs. Dens of both sorts are often in tree cavities where these are available. A favorite tree will show wear on the bark from frequent passage. Otherwise, any hollow may be used, especially if it has two entrances to facilitate escape. In built-up, areas culverts are often used as dens for this reason. Raccoons may use a den to go into semi-hibernation, or torpor, during periods of intense cold or deep, unconsolidated snow. A thaw that melts or at least firms the snow and opens water will rouse them from their sleep and allow them to resume food gathering.

Habitat

Being omnivores that are reasonably well adapted to cold, raccoons may be found anywhere there is food at any time of year except deep winter. Their one requirement seems to be denning cavities. These may be found in mature woodlands and increasingly in aging shade trees along residential streets. Their favorite hunting areas are along the shores of water bodies but their sign may be found in uplands as well, where they feed on such things as berries, birds' eggs and the nestlings themselves. It is reported that they associate human scent trails with food and will follow them to a picnic site, for instance, in the hope of finding discarded scraps.

Trackard 11 – Bobcat

The thick and rubbery pads of a bobcat's foot are a special adaptation for absorbing the shock of jumping down from heights. Furthermore, the toes themselves are laterally quite flexible. The result is a track that is more often distorted from any sort of ideal than is the case with the relatively rigid and shallowly padded feet of most canids with which it might be confused. The variability of bobcat tracks is illustrated by the various examples in figures 11.1 and 11.2 D-G. The bobcat's tracks are marvelously expressive compared to the poker-faced (or –footed) tracks of a coyote, for instance, and trying to read the cat's mind through its feet more rewarding.

Tracks

Figure 11.1 A and B show a front and a hind print recorded in mud at Brownfield, Maine. Four toes register in both. A fifth toe is vestigial and recessed far enough up the foot that it rarely appears in a print except in deep snow. Claws don't show unless the cat is accelerating or seeking traction. Normally they are retracted to the top of the foot and are hidden in fur. With toes closed together, the prints are usually around $1\frac{3}{4}$ inches across in the case of a female or otherwise small cat to $2\frac{1}{4}$ inches across for the larger male, with a variation of about $\frac{1}{4}$ inch either way. The second toe pad from the medial side tends to be slightly advanced as in Figure 11.1 B and Figure 11.2 D, E and G. This is similar to your own hand, which has an elongated second finger.

The bobcat's secondary pad is proportionately larger relative to the toe pads than is the case with canids; three toe pads would fit in it compared to about two for a coyote, for instance. Its shape varies widely with loading, but it often shows tri-lobing on the trailing (posterior) margin and flattening or even an indentation on the leading (anterior) edge. In winter both these features may be obscured by fur. A line bisecting the secondary pad longitudinally may also point off

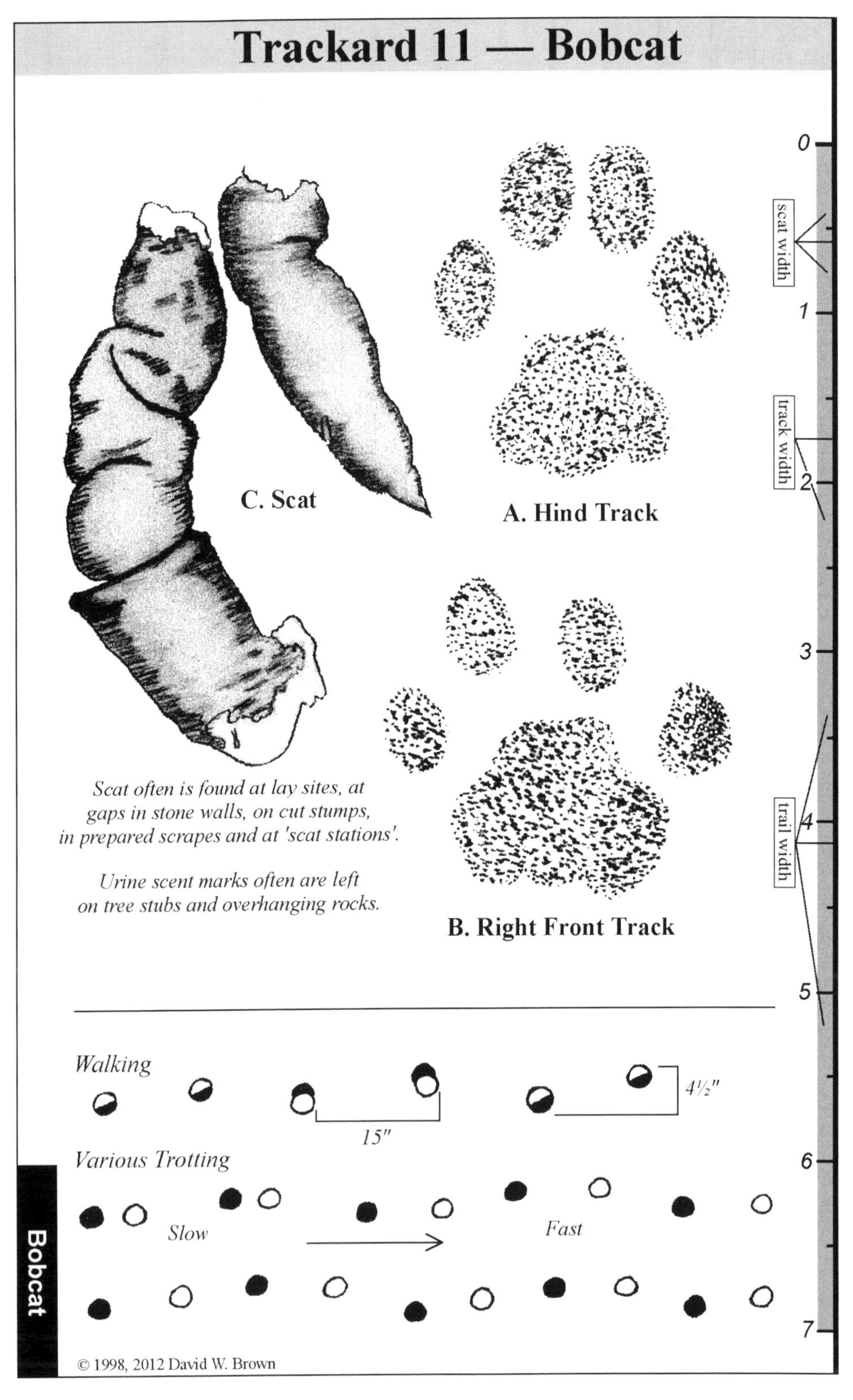

Figure 11.1. Trackard 11 – Bobcat.

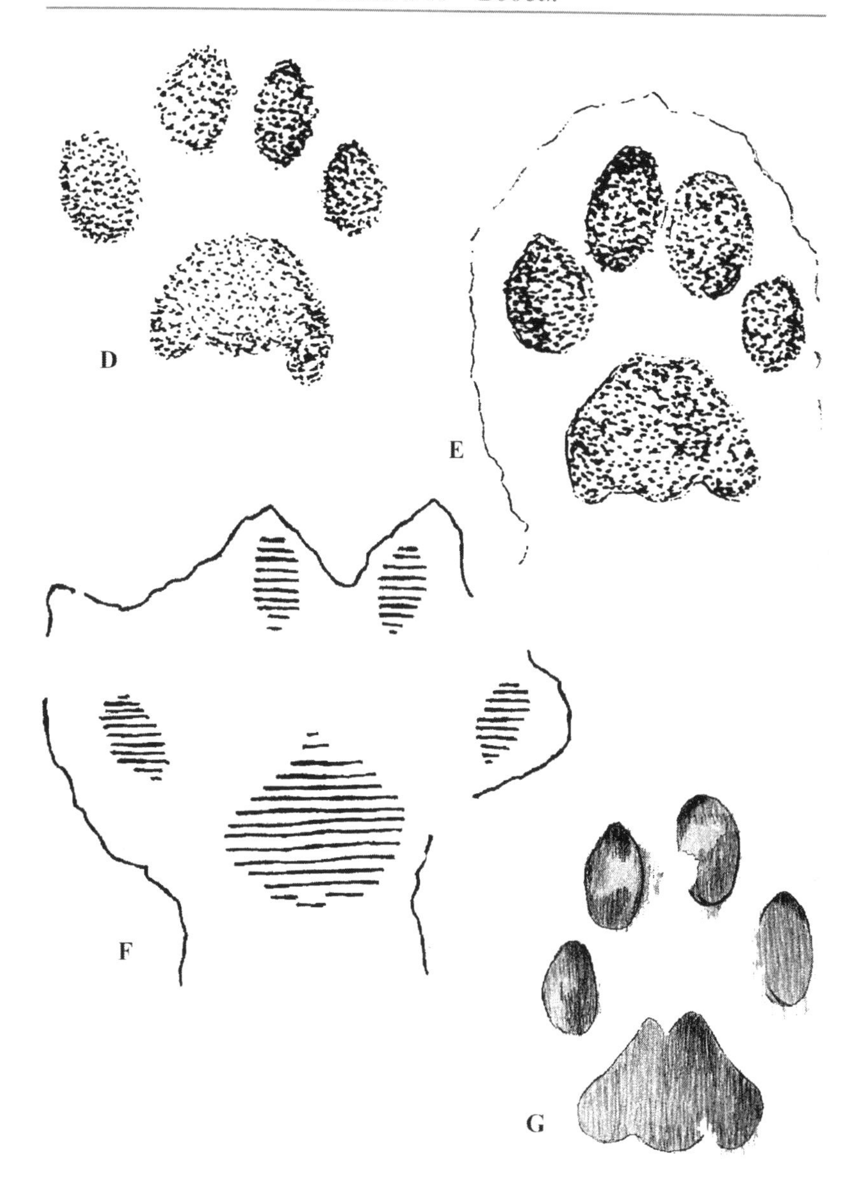

Figure 11.2. Various bobcat prints, 2/3 actual size.

D. Right front in snow. Petersham, Massachusetts.

E. Another print of same cat. Petersham, Massachusetts.

F. Spread print of female in soft spring snow. Petersham, Massachusetts.

G. Left hind of same cat in mud. Petersham, Massachusetts.

slightly toward the pad of the shorter middle toe, adding to the distinct asymmetry of many prints. In snow the area behind the arc of the toe pads will form a ridge that can be readily distinguished from the star-shaped pyramid that usually denotes dog prints. The arc of the toe pads itself is often described as being shallower than that of a dog, but on the hind print at least, this is something of an illusion created by the disproportionately large secondary pad.

Because the front feet support the head, the front toes tend to splay more than the hind and the individual pads spread out more under load, giving the impression of slightly larger front feet. The toe positions also may vary from print to print more than the hind. The secondary pad may spread under the load of the head or due to bearing the initial impact as the animal jumps over obstacles.

When front and hind register separately, the more distinctly cat-like front print generally assures an identification. But when the animal is direct registering and if the hind print dominates the appearance of the track, identification can be more challenging. This is due to the fact that the relatively lightly loaded hind tends to show a smaller secondary pad with more pronounced concave indentations on its medial and lateral sides, a more symmetrical distribution of toe pads and a slight elongation of the print in the direction of travel. The result can easily be confused with a canid print. Figures 11.1 A and 11.2 G show many of these features.

Trails

Bobcats have long and very flexible legs. They can climb trees, and they often kill by snagging fleeing prey with their sharp, retractile claws. To do either they must be able to rotate the wrist of their front feet. The result of this flexibility is an expressive trail with indirect registrations common, especially in shallow tracks.

The walking pattern of a bobcat in snow is usually the familiar alternating pattern of other walking animals. The uppermost pattern at the bottom of Trackard 11 (Figure 11.1) shows such a trail of direct and occasional indirect registrations.

The two trails at the bottom of Trackard 11 show some very common patterns for bobcat. The first is obviously an overstep walk or trot in which the front and hind feet on each side are aligned so that the hind foot passes directly over the print vacated by the front foot. The remainder of that line and the bottom line as well show a sinuous trail of single prints which is often misinterpreted as a slow loping gait. Careful examination suggests, however, that they are merely a more spread out version of the previous walk/trot pattern. With their long legs, bobcats are capable of overstepping to such a degree that the hind print disassociates itself from the front and the prints appear evenly spaced in the trail. Its relatively long trunk also permits the cat to displace slightly so that hind legs do not necessarily follow the same imaginary groove as the front. The result of all this flexibility can be a very confusing trail, indeed.

Bobcats can, of course, lope, bound and even gallop for short distances, when closing on prey for instance. But they are more careful and stealthy hunters than canids, preferring to move slowly through thick brush than to put in miles in hopes of a coincidence with prey as do the fox and coyote. In fresh soft snow a bobcat may lay up for a while until the surface firms, resting by a rabbit or hare trail while waiting patiently for prey to come into range of a leap. Or it will lie on a ledge or overhanging limb and watch the ground below for movement of cottontail or squirrel or grouse. The result is that bobcat trails are found less often than canid trails and, when they are found, they are usually walking or trotting patterns.

Despite a range that is spreading into the far North where the aggressive bobcat is reported to be displacing the gentler Canada lynx, it doesn't seem to like the sensation of loose snow on its feet. As a result its tracks are often found overlaying those of other wild animals whose trails it exploits to get around in soft conditions. The extreme adjustability of its step length allows it to walk in the prints of deer and Eastern coyote, for instance. The resulting impressions can be quite bizarre, especially if the cat was walking in the direction opposite to that of the original animal. The bobcat is also a creature of habit, revisiting the same wetland and laurel coverts time and again, sometimes even stepping in its own previous footprints in snow. Bobcat trails in winter frequently lead to southern exposures such as the brushy north edge of a powerline where the snow crusts thickly and melts early.

Track and Trail Comparisons

The identification of cat and dog tracks is complicated by several factors. It is sometimes suggested that cats show a ridge behind the toe pads rather than a central pyramid or X typical of dogs. But this ridge is not as apparent in the hind print of a bobcat and many dogs show such a ridge in the prints of their front feet. This variation in detail can be very confusing in the field when the front print of a dog or the hind print of a cat dominates the appearance of a direct registration.

Domestic dog: Dog tracks are rarely confused with those of bobcats; more often, the cat track is confused with that of the dog. Prominent nail marks and a general sloppiness of track distribution usually give the domestic animal away, while the usual lack of nail marks and a more consistent pattern of track placement show bobcat.

Gray fox: Gray foxes are rather feline canids whose tracks can be confused with those of cats. In general their prints are smaller than the usual for bobcat and, in a clear walking track, will show a much smaller secondary pad. However, in a trotting gait with a fair amount of verticality (bounce) the deep, rubbery pads of the gray fox will spread on impact, leaving an impression rather like a cat's track.

Gray fox walking trails usually show a step length of 9–13 inches, short for a bobcat. However, these foxes can leave a trotting trail of alternating direct registrations with a considerably longer step length. Even the trail width of a gray fox with its fairly wide stagger resembles that of a bobcat, and following the trail to the base of a tree where it disappears will not rule out gray fox either since this species of canid can climb trees! Obviously, then, when gray fox tracks and trails resemble bobcat, some care needs to be taken to distinguish the two. Following an ambiguous trail to a urine deposit will usually tell the difference. But be careful as sometimes gray foxes will scent mark a post previously marked by a cat, with the feline odor of ammonia overwhelming and disguising the milder scent of the gray fox.

Red fox: With a step length typically of about 15 inches and a straddle that can be as narrow as 3½ inches, the walking trail of a bobcat can look quite similar to that of a red fox. The fact that both animals have about the same size foot compounds the potential confusion. It is sometimes suggested that bobcats have a wider walking straddle than the extremely narrow stitching of a red fox. Although this is usually true, it is not totally reliable; sometimes bobcats walk finely, toward the 3½-inch end of their distribution and sometimes foxes stagger a little, toward the 4-inch extreme of their spectrum. A better clue, assuming that the distinguishing details of the prints are obscured by loose snow or whatever, is to look for the rounder impressions of the cat's foot compared to the canid's oval prints with long axis in the direction of travel. This difference can be distinguished even in old, melted-out prints, assuming that the sun worked on them evenly. Red foxes rarely show any toe drag in their trails, even in deep prints, while bobcats occasionally do. Cone-like drag impressions into or out of tracks in a walking trail point to the cat. A third bobcat feature is the tendency for the step length in direct registering walk or trot patterns to vary more widely than that of a fox or coyote. As was said, bobcats have very long and supple legs that they use to good advantage on the irregular surface of a forest floor or wetland margin, varying their steps to accommodate minor obstacles they encounter. A step length in these patterns of as much as 22 inches is possible, 7 inches longer than the normal walking pattern on a smooth even surface. Once I even found a direct register walk pattern with a 27-inch step length! As it was in dense, low pine regeneration, I attributed this to a low stalk where direct registering insured a noiseless approach.

Canada lynx: This denizen of the far North has huge feet, designed for flotation on snow, and so confusing its tracks with those of the bobcat would not seem likely. However, Figure 11.2 F shows the print of a small female bobcat on collapsible spring corn snow where the animal had spread its toes for maximum support. In such snow, pad impressions are often vague and measurement across the entire impression of snow broken by the foot yields 3¼ inches. Under these conditions a male bobcat may show a 3¾-inch track width! Note, however, that if other bobcat tracks and sign are found in the area, then a lynx identification is unlikely since bobcats will not tolerate their northern cousins in the same home range.

House cat: When a house cat is let out at night it doesn't hang around the back porch until morning waiting to be fed. Instead, it goes hunting in any nearby woods, swamps or field margins, in precisely the same habitats favored by bobcats and in precisely the same ways. I often find house cat trails in bottomlands as much as a half-mile from any house. The main differences between the two are track size and step length. A house cat's print is usually about 1¼ inches across closed toes, and its step length is about 7–10 inches. Occasionally a small or young bobcat will leave measurements at the high end of this range, and the front track width of a large tabby can reach 1½ inches. In very clear prints I have found that the secondary pad of a house cat sometimes shows a signature detail; the lateral and medial lobes may be separated from the central area by a pair of seams which run forward from the posterior indentations. This is not evident in all tracks but when it shows in a track of ambiguous size, it points to domestic cat.

The relative wildness of house cats, as opposed to the thorough domesticity of a dog, shows in their trails and adds to the difficulty of distinguishing the two felines. Followed long enough, the bobcat's trail will express leeriness toward houses, roads and other man-made features whereas the house cat, of course, is comfortable around them.

Scat

The bobcat is an habitually curious animal. It will investigate the scat left behind by other animals in its range, sometimes leaving the impression that it was the depositor. When the scat is, in fact, its own, it may be deposited at random on ground or snow or more often in a prepared scrape. It may also be used to mark the animal's passage by being left in a prominent location such a gap in a stonewall as was the case with Figure 11.1 C, from a photograph in Petersham, Massachusetts. It may or may not be covered depending on whether the animal wanted its presence or passage to be known. A bobcat may also deposit multiple scats at "scat stations," favorite sites within its range for this activity. I have found them on exposed ledges on the forested shoulders of mountains, at large isolated boulders in the woods and under powerline pylons surrounded by thick brush.

Figure 11.1 C shows "classic" bobcat scat: tubular, segmented, without large bone chips and with a maximum diameter of nearly an inch. This scat also shows a distinctive trait that I often find in the scats of this felid, a peculiar segmentation that resembles a ball and socket joint. It should be noted that segmentations sometimes appear in the scat of other carnivores, as well, and do not always appear in a bobcat's scat. Figure 11.3 H-J shows how wide a variation in a bobcat's scat appearance is possible. Figure 11.3 H shows a nearly unsegmented scat while Figure 11.3 J illustrates a case where the segmentation has been exaggerated into several discreet lumps. I attribute the former to the fact that the bobcat had had an unusually large meal from a deer carcass. In the latter instance the lumps were deposited on the surface of frozen spring snow without any effort to prepare a scrape or to cover after deposition.

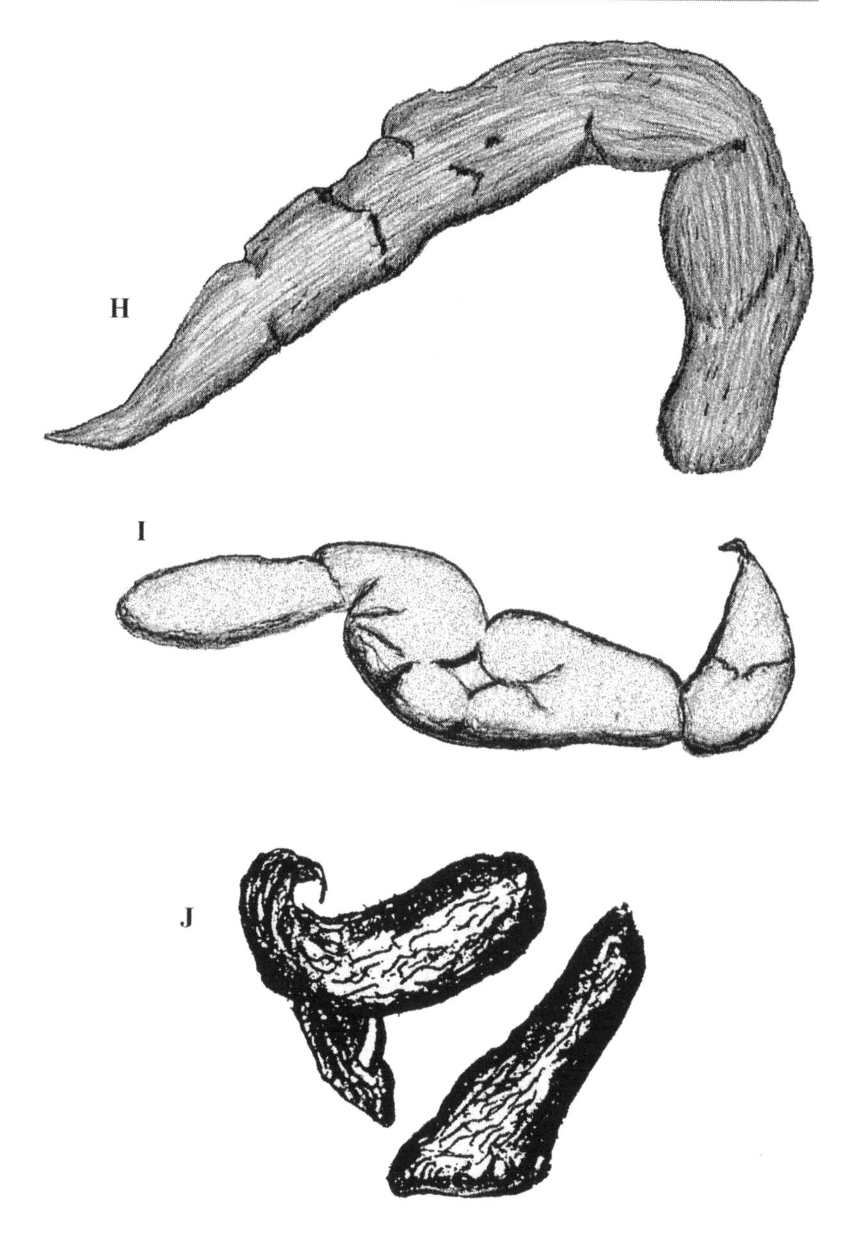

Figure 11.3. Various bobcat scats, 4/5 actual size.

H. Hair scat after having scavenged deer. Petersham, Massachusetts.

I. Hair scat deposited in prepared scrape on snow-covered stump. New Salem, Massachusetts.

J. Lumpy scat showing bird talon. Jackson, New Hampshire.

Scat from a full-size bobcat is usually larger in diameter than the scat of a fox, but may look very much like that of the Eastern coyote, which often measures in the bobcat's range of ¾ of an inch in diameter or larger. A bobcat's deposit may be distinguished by being less copious and will not contain the large bone chips that often show on the surface of coyote droppings. Bobcats do not have the teeth to crush and grind large bones, preferring instead to cut off flesh with their sharp rear teeth, called carnassials, or lick it off the bone with their raspy tongue, swallowing each mouthful whole along with whatever hair or feathers are adjacent. As a result, the consistency of bobcat scat tends to be more uniform throughout the deposit.

Certainly scat either left in a scrape or covered indicates a feline, but, as has been suggested, this feature is not always present. And the scratching that canids often perform beside or near their scat can be mistaken for preparation or covering. This scraping by canids will be beside, not under, the scat, as is usually the case with a bobcat, and may be accompanied by a urine mark as well, the smell of which will be distinctly different from the ammonia odor of cat urine.

Sign

Bobcats are aggressively territorial and often scent-mark standing objects with their urine. They may pick a class of objects for this attention, such as the dead stub of a small tree or a boulder with a slight overhang. Once one determines what sorts of things a bobcat prefers to mark, its presence can be detected by sniffing each of those objects that are encountered along a suspected route. Since bobcats are also creatures of habit as far as their hunting routes are concerned, repeatedly passing the same points such as a favorite road crossing every few days, checking stubs and boulders in the area and judging the age of the scent will keep one informed of its activities.

Just like a house cat, bobcats like to keep their claws sharp by using a scratch post. This is often a dead tree encountered on its route. A frayed patch from 1½ to 3 feet off the ground and fine splinters at the tree's base provide clues to this use.

Dens

Bobcats use certain spots repeatedly as lay sites. These are often on a ledge with a commanding view of the surroundings. In bad weather I have found them taking refuge, under a leaning rock or the butt end of a fallen tree. Any place where there is just enough room to squeeze in is especially preferred in cold weather since such small spaces allow the animal to conserve body heat.

The only long-term use of a den is for birthing and rearing kittens from March to July. These dens may be in any opening that is well-drained and far from roads and trails. Overhanging cliff ledges are sometimes used because of their inaccessibility to humans, dogs and other potential threats to their kittens.

Habitat

Bobcats hunt in dense vegetation. Their trails often follow watercourses or wetland margins where they stealthily search the rank growth on their edges for small prey. Brushy areas such as laurel thickets in the East and chaparral in the West as well as young conifer regeneration with a lot of rabbit or hare tracks are frequently visited. Bobcats use our roads and trails less that foxes do and a lot less than coyotes, normally preferring instead to follow a roadmap of their own. In deep winter, however, they may follow seldom-used plowed roads to ease passage around their hunting range. Another exception is among the dense, prickly boreal spruce and fir high on mountains where they may use hiking trails extensively as easy routes to hunting areas. And they will also walk atop a stone wall for its length, surveying the ground below them on either side although this same visual advantage can be provided by any linear raised feature such as a fallen log or an esker winding through the woods. Another habitat for bobcats is any ledgy or rocky area, including hillside boulder fields, attractive both for small prey living in the crevices and for the viewpoints, lay sites and denning possibilities such terrain affords.

Trackard 12 – Opossum

Opossums, the only widespread species of the family Didelphidae in North America, are strange animals. Being marsupials, they have an abdominal pouch like kangaroos. The young are born as fetuses smaller than the tip of one's little finger. They crawl upward through the mother's fur to this pouch, probably attracted by the smell of her milk, where they are "externally gestated" and nursed. The portability of this arrangement relieves the animal of any extended laying up for birthing and raising offspring, permitting her to continue a nomadic existence in search of food throughout her reproductive period. The opossum's marsupial and omnivorous nature probably account for its reputation among scientists as having a primitive design so effective that, like that of sharks, it has required little modification over the eons. A prehensile tail and an opposable thumb finish off the odd design, increasing the opossum's adaptability by making it as at home in trees as on the ground. When harassed, it may "play possum" by falling down, foaming at the mouth, emitting a stench and generally making itself as unpalatable as possible. After the danger has passed, it rouses itself and walks away.

Tracks

Opossums have five long toes on each foot and naked pads that have a circular pattern of grooves like a human fingerprint. These mammals are digitigrade on their front feet, walking on their toes and secondary pads. On the hind foot, a vestigial heel area generally shows in tracks, suggesting at least semi-plantigrade locomotion.

Figure 12.1 A was rubbed and shaded from a cast collected at Saugus, Massachusetts. In clear prints such as these the hind with its opposable thumb is unmistakable. However, in its usual walking pattern, the impression of this thumb is often obscured by the front print over which it registers as in Figure 12.1 C, photographed in snow at

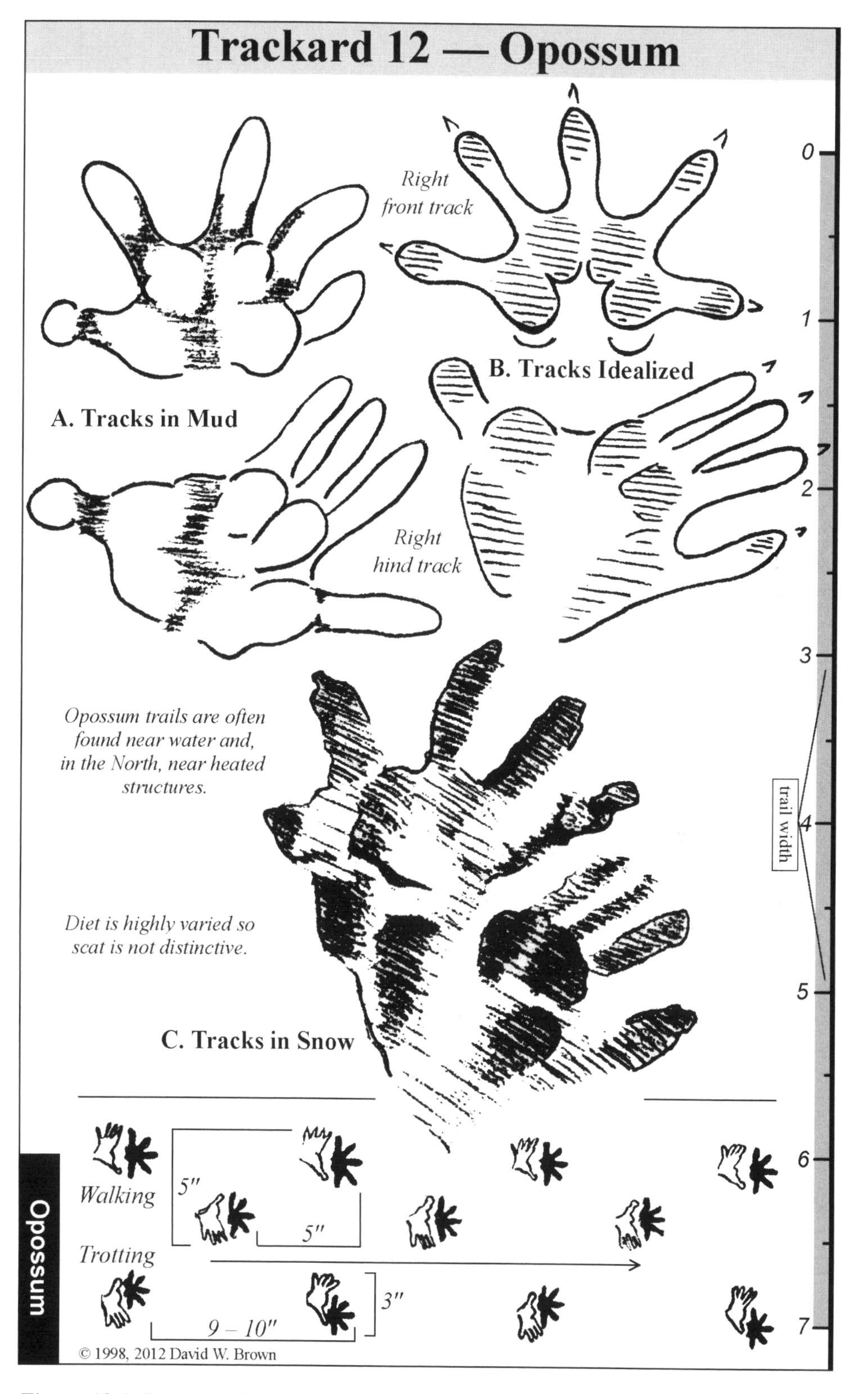

Figure 12.1. Trackard 12 – Opossum.

Lincoln, Massachusetts. Note that the other toes on the hind print are long and are commonly arranged so that the inner three are grouped together, with a distinct space between them and the lateral toe, both features that survived the double registration in this pair of tracks.

In harder conditions with shallow impressions, the distinctive print of the hind foot may register lightly or not at all. The hind feet support only the weight of the haunches and spread this lighter load over a greater area than the front feet, which support the weight of the head.

This load on the front feet may result both in the front print being deeper and in the toes of the front splaying widely, with the medial and lateral toes often pointing in opposite directions. This exaggerated splay gives the front print a distinctive star-like quality best seen in Figure 12.1 B, idealized from photographs and casts acquired on the Saugus River. This usual feature of opossum front prints shows neither in rubber casts made from the feet of dead animals nor in tracking guides which "fingerprint" dead animals' feet. A track in the wild is made by a living, weight-bearing entity under muscular control and altered by the animal's behavior as well as by the medium in which it was impressed.

In Figure 12.1 A the medial toe of the front appears bulbous like the "thumb" of the hind, but this is an illusion created by the fact that the animal articulated that toe, digging it into the mud for traction. The idealized print just to the right shows the impression of the foot when its toes are laid flat.

Track Comparisons

Raccoon: Because opossums and raccoons live in many of the same habitats, such as on steam banks and near houses, their tracks and trails are often found together. In conditions such as those described above where the distinctive hind print does not register clearly, the front print of the opossum may be confused with that of a raccoon, especially since both prints are about the same size. Although raccoons can spread their toes, they cannot do so to the extent usual on the front foot of opossums. However, if the opossum's front foot was crowded by the hind as it registered, the toes may be closed more than normal. In this case, the secondary pad holds the key to identification. This pad on the front print of an opossum is deeply segmented and symmetrical, forming the shape of a horseshoe. Raccoon secondary pads, on the other hand, are fused into a smooth surface with seams no deeper than those of the human hand. Their leading edge forms a smooth arc, and the pad is asymmetrically shaped, diminishing toward the medial side.

Trails

The usual traveling gait of an opossum is a walk, the pattern of which is represented at the bottom of Trackard 12 (Figure 12.1). This pattern is an understep walk in which the hind foot comes forward and plants just behind the front foot on the same side of the animal's body. Thus the order of tracks in the trail illustrated, moving from left to right, is left-hind left-front, right-hind right-front, left-hind left-front, and so forth. In this slow walking gait, the straddles are quite wide, with a trail width of around 5 inches, which is also the average measurement for a walking step length.

When opossums speed up to a trot, the step length spreads out to 9–10 inches and, with the stability of speed, the trail width narrows to around 3 inches. Looking casually at the two trails illustrated, one might think that the order of foot placements is the same, but a more careful look will show that in the trotting gait the pairings of feet are different. In the trot shown, the order of prints from left to right is right hind-left front, left hind-right front, right hind-left front, and so forth, that is, each two-print pattern combines diagonally opposite feet.

Scat

Opossums are nomadic omnivores that defecate at random. As a result their scat is not distinctive in shape or location. Since they are about the size of a house cat, any scat of appropriate size and undistinguished shape may be that of this animal, whatever its contents.

Dens

The marsupial anatomy of the female allows opossums to move continually in the direction of food. Even when the young are too big to fit into her pouch, the mother opossum will still carry them on her back where they hang onto her fur. Opossums will den for a while in bad weather in any convenient lodging, from woodchuck burrow to house cellar or garage. In order to insulate a ground burrow and elude detection by predators, an opossum, like other mammals that take shelter in holes in winter, may plug the entrance with dead leaves scraped together from the ground nearby.

Habitat

The range of the opossum has been spreading northward for many years, facilitated by the atmospheric warming caused by urban heat islands on the large scale and central house heating on the small. With naked tail and poorly furred ears, opossums are poorly adapted to cold and so their trails in northern regions are more likely to be found in suburban warmth than in the cold forest beyond. An appetite for any edible substance including garbage reinforces this tendency. Unlike the modest moral center of Walt Kelly's famous comic strip, opossums in nature are ill-tempered and, if discovered in your garage, are best left alone. When the food runs out, the animal will move on.

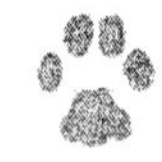

Trackard 13 – Woodchuck

Woodchucks are the eastern version of western marmots. Although they may not look like squirrels, they are actually large members of that family. The difference in their appearance is due to the need to put on a great deal of body fat during the warmer months to carry them through a long and profound hibernation. Although the western marmots are larger than the woodchuck, their track morphology is similar. Identifying tracks of the several species, then, is a matter of print-size as well as range and habitat preference.

Woodchucks live in elaborate tunnel systems that provide both hibernacula and security, and from which they make forays into fields and gardens in search of succulent growth. These tunnels are often appropriated by other animals such as foxes and coyotes for their own purposes. Other animals, such as cottontails, frequently take shelter within their entrances during bad weather.

Tracks

Because these mammals sleep through the winter, woodchuck tracks are not often found in that season. A spring snowstorm, however, may catch them abroad, especially in a woodline at the edge of an agricultural field. Figure 13.1 A, for instance, was photographed just inside a woodline in a dusting of snow over ice in late March at Lincoln, Massachusetts. In general, woodchuck prints look like those of a giant squirrel, with four toes on the front and five on the hind.

A woodchuck's front print shows four toe pads that are frequently splayed, making the front foot seem larger than the hind. These pads often have the odd arrangement shown in Figure 13.1 B, idealized from a cast in mud at Belmont, Massachusetts, as if windswept toward the medial side. The secondary pads are segmented and symmetrical. At the bottom of the print, tertiary pads usually register in an area shaped like the heel of a human hand. The tiny pad on the medial side of the tertiaries is all that is left of the fifth toe, almost

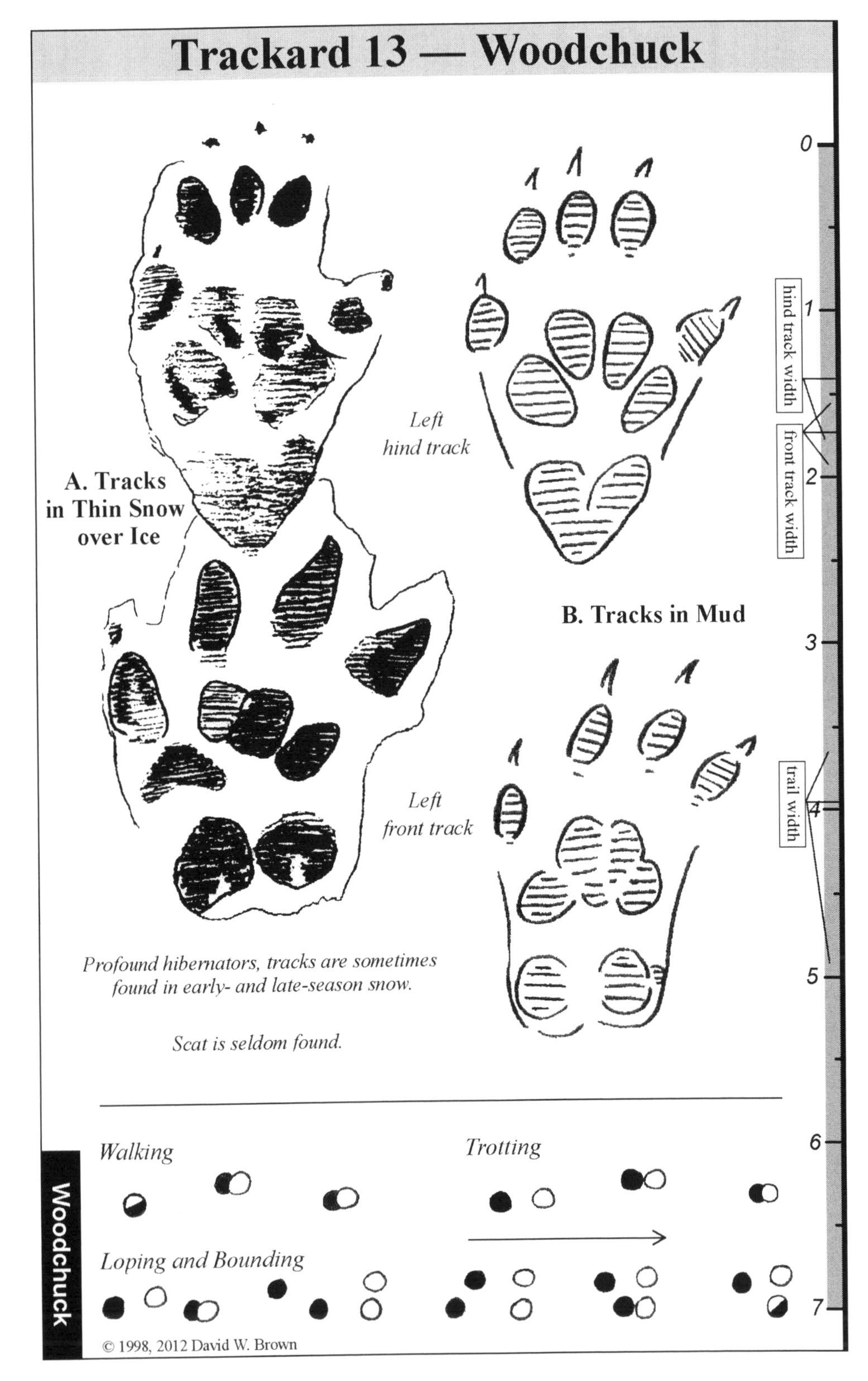

Figure 13.1. Trackard 13 – Woodchuck.

completely lost in the course of evolution. It is included in the idealized illustration but only rarely can be seen in an actual print.

The hind print shows five toes, with the middle three usually grouped and even, another characteristic of squirrels. The outside toes on either side are somewhat shorter then the middle ones with the medial toe often splayed at a fairly radical angle toward the centerline of the animal's body. The secondary pads appear as a group of four segments in a horseshoe arrangement. The heel (tertiary) area may or may not show. When it does, it is generally a diffuse impression.

Track Comparisons

Raccoon: Raccoons leave prints about the same size as those of woodchucks and, when their toes are drawn back, can show a superficial resemblance to woodchuck prints. However, raccoons have five toes on the front as well as the hind and show a fused, asymmetrical secondary pad with a smooth leading edge.

Trails

Lacking both the need to traverse snow and the gait efficiency that such travel implies, woodchucks tend to be sloppy walkers, showing frequent indirect and double registrations in their patterns as illustrated at the bottom of the card (Figure 13.1). An overstep walk or trot is most common. However, when escaping to a plunge hole in a field, for instance, a woodchuck is perfectly capable of a loping gait or any of the other faster gaits illustrated. Because of its long and pudgy body, the hind feet don't overstep the front by much, except in its fastest gaits, so the resulting patterns tend to be characteristically short.

Scat

Woodchucks seem most often to defecate underground so their scat is seldom discovered without digging. A dropping deposited on the surface will be a nondescript lump or aggregation of pellets depending on the level of succulence of its current feed. Maximum diameter is around ¾ inch. Other herbivores of similar size with which it might be confused, like porcupines, have smaller and more distinctively shaped scat. Location in agricultural or other fields in summer or along woodlines abutting these fields is circumstantial evidence for this species as is placement near feeding sign as well as near a plunge hole or den.

Sign

Woodchucks have a tooth arrangement similar to other squirrels, that is, two long, sharp central incisors, then a wide gap where human bicuspids are located, followed by a set of molars in the back of the mouth. Like other rodents, as well as rabbits and hares, they usually make a sharp, angular cut on a stem with the incisors and then feed the cutting into their mouths to be chewed by the molars. Although they may be forced by hunger to eat woody material in the early spring, they prefer green succulents such as, unfortunately, those found in gardens. Often seen in grassy areas, they are usually there not to feed on grass, which in summer is abrasive on their teeth, but rather to feed on clover, dandelions and other such plants that grow with the grass. On these plants is where cutting sign should be looked for.

Sign Comparisons

Deer: In April, deer as well as woodchucks will graze on grass in fields while it is still succulent and has little abrasive content. Deer feeding sign will look ragged while woodchuck will be neater, and deer droppings will usually be found near their feeding sign.

Other rodents and lagomorphs: Porcupines and rabbits visit the same fields as woodchucks in the spring to graze on new grass, and since neither species takes any care to hide its scat, cutting sign found there may be mistakenly attributed only to them. All three species have similar incisors and so there is little to distinguish their work at this season. Later in the summer a preference for less woody and abrasive food may help distinguish woodchuck scat. Also in early spring when there is little else to eat, woodchucks may skin off the tender bark at the base of woody shoots. This may be mistaken for similar work by rabbits. However, rabbits usually bite off some of the wood under the bark as well, leaving a rather ragged appearance. Woodchuck debarking is more precisely limited to the nutritious surface layers.

Dens

Summer tunnel systems often extend from a woodline out into fields. As these openings are danger areas for this animal, it frequently digs a plunge hole into which it may escape from the fox, the coyote, the irate gardener or the farmer with a gun. This hole is very discreet, usually in tall grass, and is dug from the inside out so that there is no fan of spoil to attract attention. Other entrances, however, usually show such a fan in front of an opening about 6 inches in diameter. Profound hibernation usually takes place in an underground chamber in a tunnel system, sometimes at a distance from fields.

Den Comparisons

Red fox and **Eastern coyote:** Both fox and coyote trails may lead to the entrance of a woodchuck tunnel, but this is not necessarily a sign that they have taken over the burrow for their own purposes. During a storm rabbits and other prey often take shelter just inside the entrance. Their predators know this and routinely investigate these openings. However, around mid-winter the females of both canid species do often take possession of a woodchuck tunnel for birthing and raising pups. But they must enlarge the entrance, to at least 8 inches in diameter in the case of the fox and more than a foot for coyotes. If a fox or coyote has weaned pups within, the spoil fan will show parts of the larger prey that they are being fed.

Habitat

Although woodchucks are sometimes seen in woodlands and can, like other squirrels, climb trees, their bulk is better adapted to terrestrial browsing in fields and gardens. Grassy areas, wherever they are found, are attractive as is succulent plant growth at the edges of wetlands. Old barn foundations on abandoned farms are favored as well; the crevices in the stones provide refuge and access to the overgrown fields around.

Trackard 14 – Red Fox

The red fox's range includes nearly all of the United States except the extreme Southwest where it is replaced by the gray and kit foxes. Although it has several color variations, the most common is the one that gives the species its name. Whatever the appearance of its fur, all red fox variations have a white tip on the tail.

In summer a red fox looks very skinny, with a tail or "brush" nearly as big as its body. With the onset of cold weather, however, this fox grows a dense undercoat of wooly fur, making the animal appear much more robust than it really is. This underfur is white, that is, hollow and translucent, making it superior insulation. Another defense against the cold is the fur on the underside of its feet, which protects against conductive heat loss to the snow. Curled up, with its brush over its feet and nose, the red fox can withstand very cold temperatures without moving a muscle. Able to detect voles under crust or deep snow with its incredibly acute senses, the red fox, then, is admirably adapted to the winter environment.

Tracks

Figure 14.1 A and B show front and hind prints of an adult in muddy sand at Brownfield, Maine, in May. Both are fur-covered and remain so in summer as well as winter. In very clear mud tracks, the fleshy toe pads may show through the fur as small ovals. But in other conditions, such as dry, loose snow, these stiff hairs result in very diffuse prints without sharply defined pads. In gaits faster than a walk, front and hind prints appear separately. However, in a normal walking pattern, the hind prints are impressed so neatly within the outline of the front (direct registration) that they look like the result of a single foot placement. Both prints are somewhat longer than wide and, like the coyote, the red fox normally keeps its toes together, leaving a very tight print. However, when it accelerates to high speed, it will not only spread its toes but also dig in its usually inconspicuous

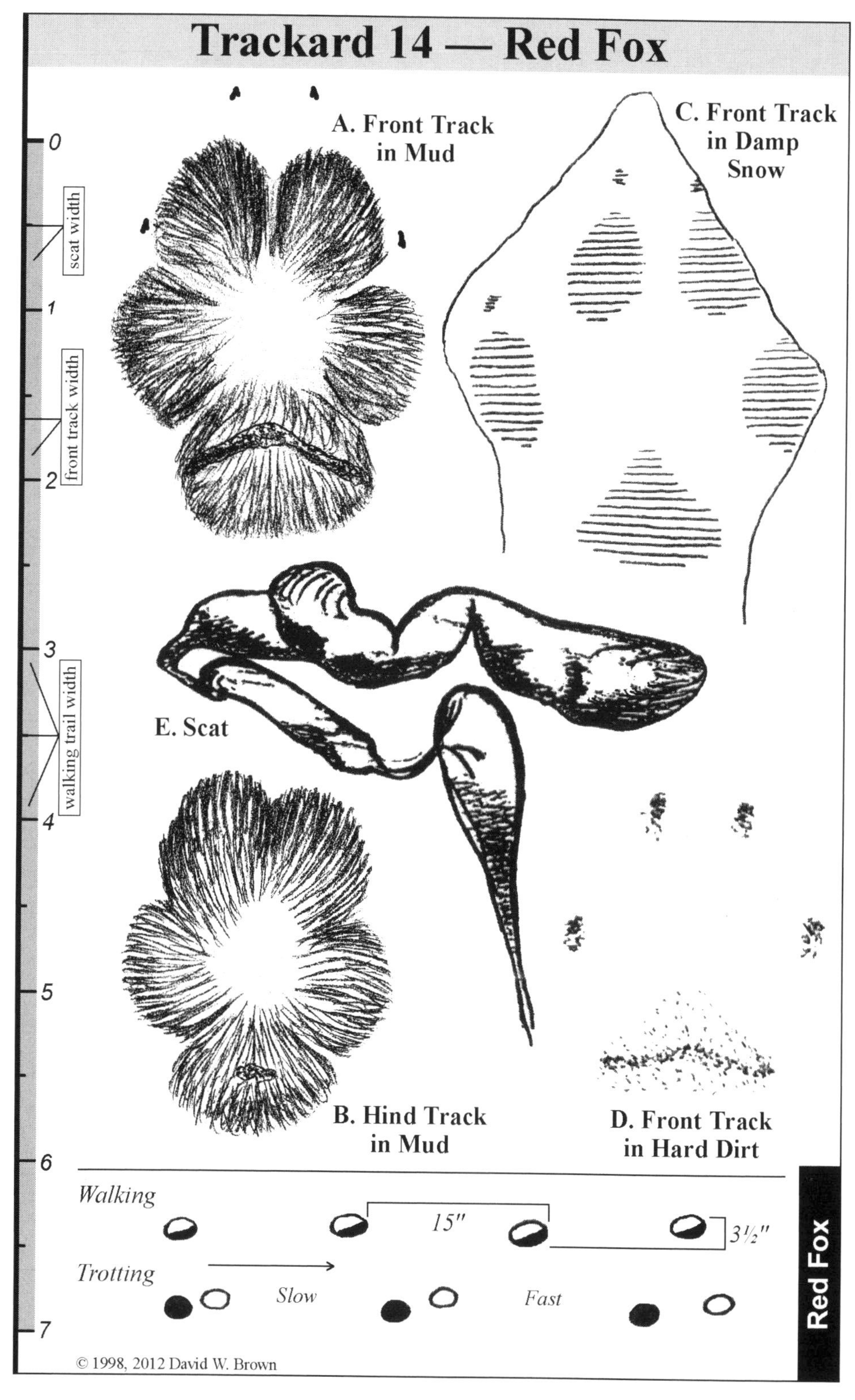

Figure 14.1. Trackard 14 – Red Fox.

nails as in Figure 14.2 H. This is true as well in other situations where the fox needs traction or support. In deep, wet snow the fox often spreads its toes to a track width of as much as 2½ inches, ¾ inch wider than normal, to keep from collapsing through the surface as in Figure 14.1 C, photographed in late winter corn snow at Lincoln, Massachusetts. In these conditions a red fox track can look very much like the splay-toed track of a domestic dog. The neat direct registration, however, as well as trail characteristics detailed below, should easily distinguish them.

The front print of an adult red fox normally measures about 1¾ inches across the toes when they are closed. Young foxes or small vixens may measure as narrow as 1½ inches as in Figure 14.2 F, showing the smaller hind print of a young red fox in August of its first year. At the other end of the statistical distribution, a closed-toe print of a large adult in soft conditions may occasionally measure a full 2 inches across.

Since they are used for digging, the nails on the front feet are longer than on the hind. The side nails are buried in fur and so are usually inconspicuous, at least in slower gaits and on firm surfaces. They register surprisingly far forward, laterally bisecting the middle toe pads.

The width of the front secondary pad of an average adult will measure about 1⅛ inches and will always be the same width or narrower than the middle two toe pads. This is important in distinguishing red fox from a small coyote, whose front secondary pad in usual closed-toe prints will be wider than the middle two toes. This fact makes normal red fox front prints and direct registrations both laterally and longitudinally symmetrical. If you come upon the trail of a canid and are not sure at first which way it was going, chances are you are looking at a red fox's trail.

The most distinctive feature of a red fox's front print is a bar or "chevron" of naked callous that runs all the way across the width of the otherwise fur-covered secondary pad. This bar appears in all but the vaguest prints and is distinctive to this species. Figure 14.1 D was photographed in Templeton, Massachusetts, on a sandy road that was so hard that the pads barely registered. Nevertheless the signature bar was detectable and led first to the discovery of the print and then to the faintest of trails.

The hind nails are shorter and less conspicuous in the print than are the front nails, often not appearing at all. The hind foot is also

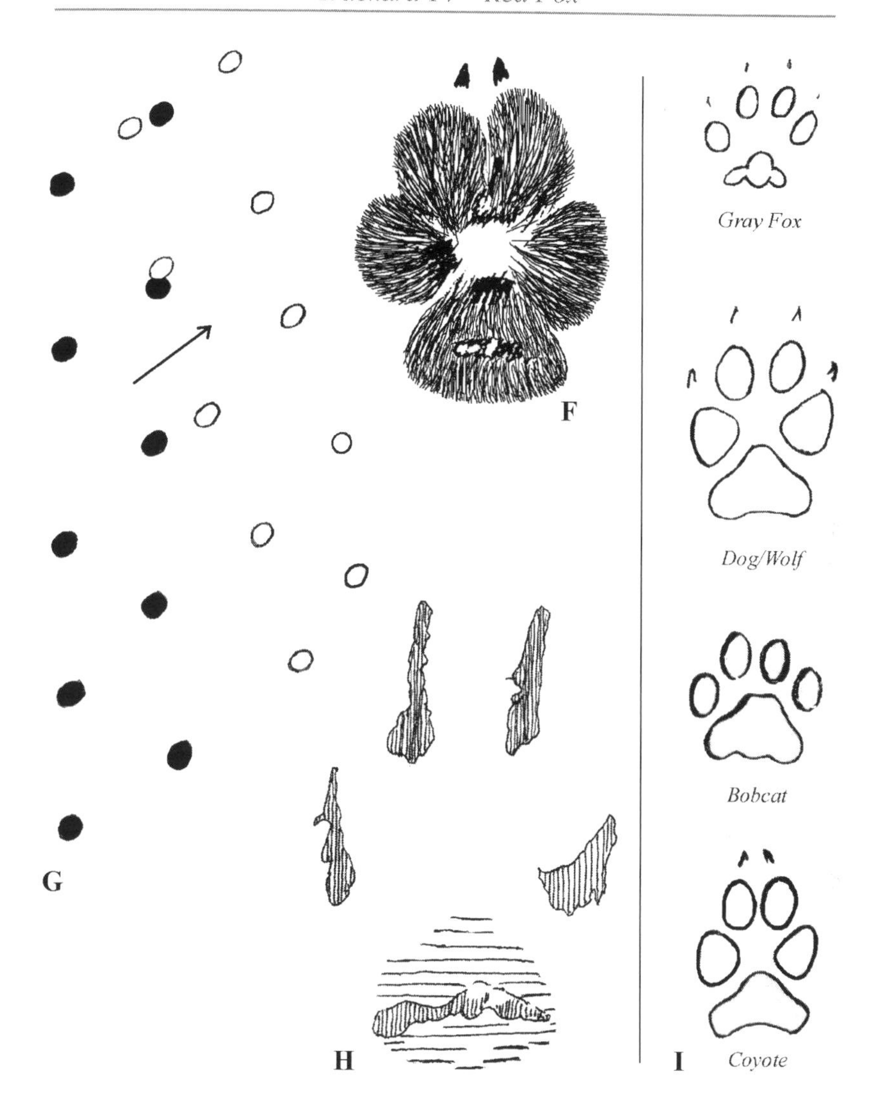

Figure 14.2. Red fox tracks and patterns.

F. Hind print of young fox in mud, 4/5 actual size. Brownfield, Maine.

G. Various lope and gallop patterns with increasing speed from top down.

H. Print of a galloping adult in muddy sand showing nail marks for traction. Templeton, Massachusetts.

I. Profiles of similar species for comparison, greatly reduced.

hairy and, as a result, leaves a print even more diffuse in dry snow than the front because, with only the weight of the hindquarters to support, the rear foot impresses more lightly. There is usually a small calloused area exposed in the center of the secondary pad but it is not distinct. In fact the whole pad usually shows as just a vague circular area without its lateral lobes. Even when the hind print direct registers over the front in a walk, it is so light that the underlying front print still usually dominates the appearance of the track.

Trails

The walking pattern of a red fox is illustrated at the bottom of Trackard 14 (Figure 14.1). In snow or sand it is distinctive for its narrowness and the perfection with which each hind print registers directly over the front, leaving the illusion of a single print. Each of these front/hind tracks registers almost directly in line with the last so that a fox walking across firm snow on a frozen pond, for instance, leaves a trail as narrow and regular as sewing machine stitches, with an average in New England of 15-inch step lengths and a trail width only an inch or so wider than the tracks themselves. A narrow, direct registering trail is much more efficient in snow or other deep footing than is the wider, indirect registering trail of a domestic dog, for instance, and displays an instinct tutored by millennia of ruthless selection.

On a firm, even surface, the usual traveling gait of a red fox as well as other canids is a "displaced trot," a gait in which the animal skews its spine to one side or the other so that its hind feet can pass by the front. In this pattern, illustrated as well at the bottom of Trackard 14 with increasing speed from left to right, all the front prints are lined up on one side of the trail and retarded while all the hind prints are on the other side of the trail and advanced. This gait is common with coyotes and domestic dogs as well.

Red foxes also use a variety of lopes for faster traveling and can even gallop for short distances, a gait generally reserved for escape. Figure 14.2 G shows a series of loping and galloping patterns with increasing speed from top to bottom. So narrow is the trail of a red fox at speed that the bottom two patterns, representing gallops, may appear at first glance to be the same. Careful examination will show that the upper of the two is a "transverse" pattern, that is, the prints alternate in a placement order, from lower left to upper right, of front-right, front-left, hind-right, hind-left. The gallop pattern at the extreme bottom illustrates, a "rotary" pattern with the order front-left, front-right, hind-right, hind-left. Red foxes mix these two basic patterns at will.

Track and Trail Comparisons

Bobcat: The track width and walking step length of bobcats and red foxes can be quite similar. However, even a vague bobcat track looks round compared to the red fox's diamond, oval or arrowhead shape with the long axis in the direction of travel (see track profile comparisons of bobcat, as well as several other species in Figure 14.2 I). Also where a red fox, like other canids, leaves a mound or pyramid in the center of its prints, cats leave an arc or lateral ridge, especially in a front print, which is particularly conspicuous in snow. Toe arrangement in a bobcat's print is usually somewhat asymmetrical and, finally, a cat's claws will not show in a print unless it had some need for traction.

Not only are bobcats' feet similar in size to those of red fox, but they often walk with the same 15-inch step length, as well. However, in firm conditions bobcats indirect register more often than do red foxes and also tend to vary their step length more in uneven terrain. Finally, a walking bobcat often shows a wider trail width than a red fox, giving its trail a slightly more zigzag appearance.

Coyote: Coyotes have naked pads that register well-defined edges in most conditions. Their secondary pad lacks a bar and the front secondary pad is usually wider than the middle two toe pads, at least in normal closed-toe prints. Coyotes, especially the eastern variety, normally leave larger prints, but December tracks of young animals may overlap with expected measurements for red fox. Red foxes are light animals with big feet compared to coyotes, which are larger but with smaller feet for their size. Thus in similar soft conditions, foxes will leave a rather light, delicate print while coyotes will leave a deeper and more robust impression.

A young Eastern coyote in December or January, before it has reached full growth, may walk with a step length as short as 17–18 inches, 5–6 inches shorter than a typical adult. Given the smaller print size of such a juvenile there can be some confusion with red fox. Following such a trail for a while will show the indications of its immaturity, however, with dashes to one side or the other to satisfy curiosity at the expense of energy. A large red fox at this season will be an experienced adult and its trail will express mature sobriety and efficiency. At this season, as well, coyotes often travel in family "packs" of several animals following one another through the forest. Red foxes are more solitary animals in winter; one may occasionally see the trails of a mated pair consorting, but usually foxes travel alone.

Gray fox: Gray foxes are slightly smaller than reds and their front prints are rounder, catlike and comparatively hairless. The toes on front prints tend to splay even in fairly firm conditions, and the secondary pads on front and hind are smaller, lack a bar and often show a distinctive "winged ball" shape quite different from the vague, hairy impression typical of red fox.

Gray foxes are less well adapted to northern winters, having only recently recovered the northern part of their traditional range. As a result, their walking trails often show inefficiencies that a well-adapted adult red fox would never be guilty of, such as occasional or sometimes regular indirect registrations, especially on firm surfaces. The typical walking step length is also shorter than that of a red fox, varying generally between 9 and 13 inches depending on the size of the animal. In soft snow, however, the gray may stretch its walking step to 15 inches and may even occasionally leave a trot pattern of alternating direct registrations up to 20 inches long. In most walking trails, however, gray foxes show a trail width with a little more stagger than the almost straight stitching of a red fox. This is not always so but is true often enough to serve as a clue that may be added to other evidence.

Domestic dog: A small domestic dog may leave fox-size prints but with few of the characteristics mentioned as typical of red fox. In addition, even lap dogs are descended from wolves and typically show wolf-like splaying of the toes even on firm surfaces; except on collapsible surfaces, red foxes maintain a fairly tight pad arrangement. Since domestic dogs don't dig for a living, their nails will not be worn down and, unless they have been clipped, will show prominently in their prints.

A fox-size domestic dog's trail will show the effects of its domesticity: foot drag, wide straddle, indirect and double registrations and a tendency to dash and romp excitedly wherever its nose leads.

Scat

Classic red fox scat is pictured, actual size, in Figure 14.1 E from Saugus, Massachusetts, in February. This scat, with a maximum diameter of a little more than ½ inch, is composed of the fur of the fox's prey. The "tail," which may appear at one or both ends, tells only that the dropping came from a carnivore since the tail is actually the excreted hair or feathers of prey.

In late summer red fox scat may be composed entirely of seeds and pits, giving it the appearance in Figure 14.3 K. In such cases it may look like scat of any other berry-eating mammal of similar size. The tendency of foxes, as well as other predators to deposit their scats on prominent sites as markers should help to distinguish them from non-predator berry-eaters.

Either for vitamins, as some have suggested, or to excite bowel movement in an animal that relies mostly on a binding diet of muscle and fat, red foxes often eat fibrous material. Figure 14.3 L shows a grass scat that was left in the middle of an unused dirt road in Petersham, Massachusetts. In this case the "tails" are simply grass fibers rather than the fur of prey. I have occasionally found red fox scat composed of pine needles and even the stuffing from a discarded automobile seat. Apparently anything that produces the desired effect will do.

Red foxes cache uneaten prey a lot. In winter a fox that has failed to meet with hunting success may dig up an old cache buried in the snow. As the carcass is reduced, more fur and feathers than muscle will be consumed, resulting in a dry, compact yellowish-brown scat similar in appearance to that of coyotes and other predators that use a similar tactic. Figure 14.3 M shows an example of this dry winter scat.

Scat Comparisons

Eastern coyote: The scat of this canid is often similar to that of red fox both in composition and location. Generally, Eastern coyote scat will have a diameter of ¾ inch or more and may show large bone chips on its surface, a result of wolf-like jaws that are much stronger than those of a fox. However, sometimes red foxes seem to get constipated and leave lumps of scat thicker than normal. In these cases the amount of scat is a better indication of species. Red foxes have very small stomachs and so feed only a little at a time. The resulting scat reflects this small capacity compared to the normally copious scat of an Eastern coyote.

Domestic dog: Dogs also leave copious scat, but with a grainy consistency and repulsive odor. The carnivore scat of a wild predator, on the other hand, has an almost sweet odor that is quite different.

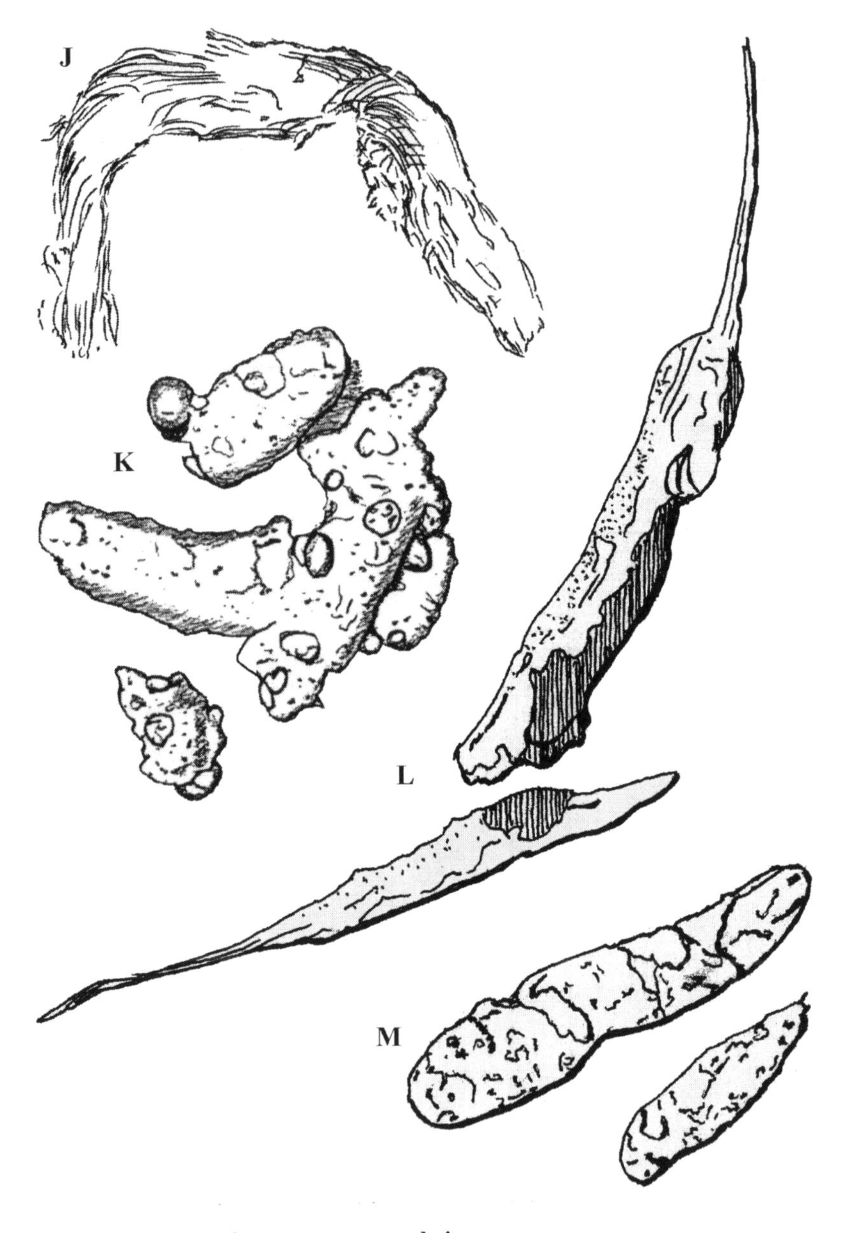

Figure 14.3. Red fox scats, 4/5 actual size.

J. Scat composed entirely of cottontail fur. Concord, Massachusetts.

K. Berry scat on an exposed rock in a dirt road. Brownfield, Maine.

L. Grass scat. Petersham, Massachusetts.

M. Dry winter scat. New Salem, Massachusetts.

Warning: Be careful with the scat or urine of any animal. Do not handle it or otherwise come in contact with it as serious illness can result. Be especially careful fractionating dry scat because invisible spores may become air-borne and get inhaled inadvertently.

Bobcat: Bobcat scat usually has a maximum diameter larger than red fox, but I have occasionally found very small and narrow deposits by the cat. Sometimes, but not always, bobcat scat shows characteristic segmentations and may actually break into lumps, a characteristic not common in foxes. Classic red fox scat such as the one on Trackard 14 may show stringy constrictions, a trait I have not found in bobcats. However, the sum of the evidence upon which most identifications depend should include not only appearance but also the context in which the scat was found. Deposition in a prepared scrape announces a feline, keeping the caveat in mind that on a hard surface like a rock or a hard road the scrape may be absent. Different predators tend to deposit their scat on a particular class of objects, as well. Although I often find red fox scat on or along trails, I seldom find the same with the bobcat, which tends to cross rather than follow them. Instead, the bobcat seems to prefer stone walls or their gaps as well as snow-covered stumps, places that red foxes in my area, at least, don't favor.

Fisher and **Gray fox:** See the illustrations of these species' scats in the appropriate chapters (figures 6.1, 6.3, 16.1, 16.2). They can be similar to those of red fox and the caveats expressed above apply. Both of these animals, but especially fishers, leave deposits on snow-covered stumps like a bobcat often does but without the prepared scrape. Gray foxes, fishers and bobcats are adept at preying on squirrels, which they can pursue up into trees. A fractionated scat that shows squirrel fur and bones is more likely to be from one of them, therefore, rather than from the terrestrial red fox, whose favored prey animals are rabbits and voles.

Scent Marking

Red foxes have several scent glands on their bodies, but the scent most often detected by humans is the urine mark. The odor is often characterized as "sweet skunk." It can be distinguished from skunk by its extremely localized effect. Whereas the odor of skunk spray will spread over a large area, red fox scent is often detected by the passer-by as just a whiff that disappears in a step or two. These scent marks are commonly placed along trails where the fox may select a class of objects to anoint such as fir seedlings of a certain height. In mid-winter when mating is taking place, red foxes seem to urinate in greater or more concentrated amounts so that on damp, still days the odor may fill the woods almost on the scale of a skunk discharge.

By way of comparison, the urine scent marks of fisher and gray fox usually are lightly scented and smell to the author's nose like chemically treated paperboard, not like skunk. Coyote urine also has a skunk-like odor but tends to be more copious and less pungent than red fox. It also has the odor of burnt hair mixed in, an odor you may have smelled from your hair dryer. The rule of thumb I use is that if I can smell it standing up, it's a red fox; if I have to kneel down to detect it, it's a coyote. Domestic dog urine does not have a musky odor.

Dens

Red foxes use dens from early spring to mid-summer for bearing and raising pups. At other seasons they may investigate entrances to burrows during storms looking for prey animals that use them for shelter. However, red foxes are so overbuilt for cold that they rarely rely on dens for protection in the winter themselves. Around December vixens start looking at potential birthing dens in their range, renovating several of them that they then may use as centers of their hunting activity. Sentinel lay sites in the snow around the entrances may be found where the fox sunned and perhaps consumed larger prey. Whelping occurs in early spring after which the vixen may use more than one of the prepared dens for rearing, transferring her kits if she is disturbed by humans or dogs, if the current burrow gets too hot at the approach of summer or if the den becomes infested with vermin. Red foxes often take over a woodchuck burrow and renovate it for their needs, enlarging the entrance to at least 8 inches in diameter. These are usually on slopes with good drainage, and spring dens, at least, often have a southern exposure. After the kits begin to emerge from the birthing den around April/May, the spoil fan at the entrance will show the remains of larger prey such as rabbits, opossums, muskrats, raccoons and ducks. Smaller staple prey such as voles and mice get consumed entirely, and their remains can only be found in scats.

By way of comparison woodchuck den entrances are normally about 6 inches in diameter, a couple of inches narrower than the minimum for a tunnel renovated by a red fox. Woodchuck burrows appropriated by Eastern coyotes will have entrances more than a foot in diameter. Although gray foxes sometimes use an elevated den in a tree hole, a ground den will look just like that of a red fox. However, it will not have a musk odor clinging to it. (Be aware that skunks sometimes take over old dens and leave their odor clinging to it.) Squirrel remains at the entrance bias toward the more arboreal gray fox.

Habitat

The classic habitat for red foxes is a pastoral landscape of mixed fields and woods that provides a lot of edge brush and hedgerows for hunting and den concealment. However, these foxes are highly adaptable in both their social arrangements and habitat preferences. They can do quite well in deep forest and high mountain environments as well as in suburban parklands and even in cities. The red fox's dense winter underfur, fur-covered paws and huge brush with which it can cover its extremities when at rest permit it to range as far north as the lower Arctic. Many kit foxes of the Southwestern desert, interestingly, have retained the stiff fur covering their pads, perhaps as a useful vestige of a boreal ancestry, not to prevent conductive heat loss to snow, as with reds, but rather to protect their pads from the searing desert sands.

There is a long-standing myth about the origin of North American red foxes. The story has it that our red foxes were originally imported from England by Virginia gentlemen for fox hunts, having found the native gray fox unsuitable as its talent for climbing trees made it easy to throw off the hounds. The red fox's thorough adaptation for cold, however, suggests a northern origin, perhaps migration into North America over the Bering Land Bridge from Siberia, followed by dispersion southward over the continent. Perhaps the most compelling evidence for red fox presence pre-dating European colonization, however, is that Native Americans had words for both red and gray foxes in their vocabulary. While prowling the stacks in the Concord Free Library years ago, I found a book written by Roger Williams, the founder of Providence, in which he documented the language of the Narragansett Indians, whom he encountered early in the 1600s. In it he listed words for both red and gray fox. The red fox was "misquashim" and the gray "pisquashish." As neither word appears to be derivative from the languages of any earlier European visitors and since the Indians would not have had a word for something they didn't know, it seems certain to me that the red fox was here before the white man.

Trackard 15 – Muskrat

Muskrats are rodents adapted to watery environments. They have brown fur, long, thin naked tails and a body shape similar to that of their larger cousin, the beaver. In habitats where both reside and at distances where size is hard to judge, these two species can be told apart by a number of field marks. When they are swimming, muskrats sometimes raise their narrow tails in an arc over the water while beavers keep their large tails flat on the surface or submerged. Beavers have fleshy exposed ears while those of muskrats are covered by fur and are inconspicuous. Finally beavers have wide, webbed hind feet that allow them to cruise their ponds with the sedate air of proprietorship, their slow leg turnover creating scarcely a ripple. Muskrats, on the other hand, have much smaller feet that they must paddle furiously to make headway. The result is a lot of rippling around their bodies that can be seen in the wake they leave behind. Beavers are fat, complacent landlords while muskrats are harried tenants always rushing to make ends meet.

Tracks

Muskrats have five toes on each foot. The front print is much smaller than the hind (Figure 15.1) and in a commotion of tracks on a streambank, this difference and the presence of a tail mark often lead to the discovery and identification of the species. On both front and hind feet, muskrats have long nails, a feature that is sometimes disguised when the animal digs its nails straight down for traction on slippery mud as in the profiles shown in Figure 15.2 E.

When the animal is moving slowly, with its weight balanced on its hind feet, front prints may register lightly and be hard to see. Sometimes as well they are covered by the larger hind prints. Four toes are prominent, with the medial fifth toe extending off the tertiary (heel) pad area and vestigial, although more pronounced than in other rodents.

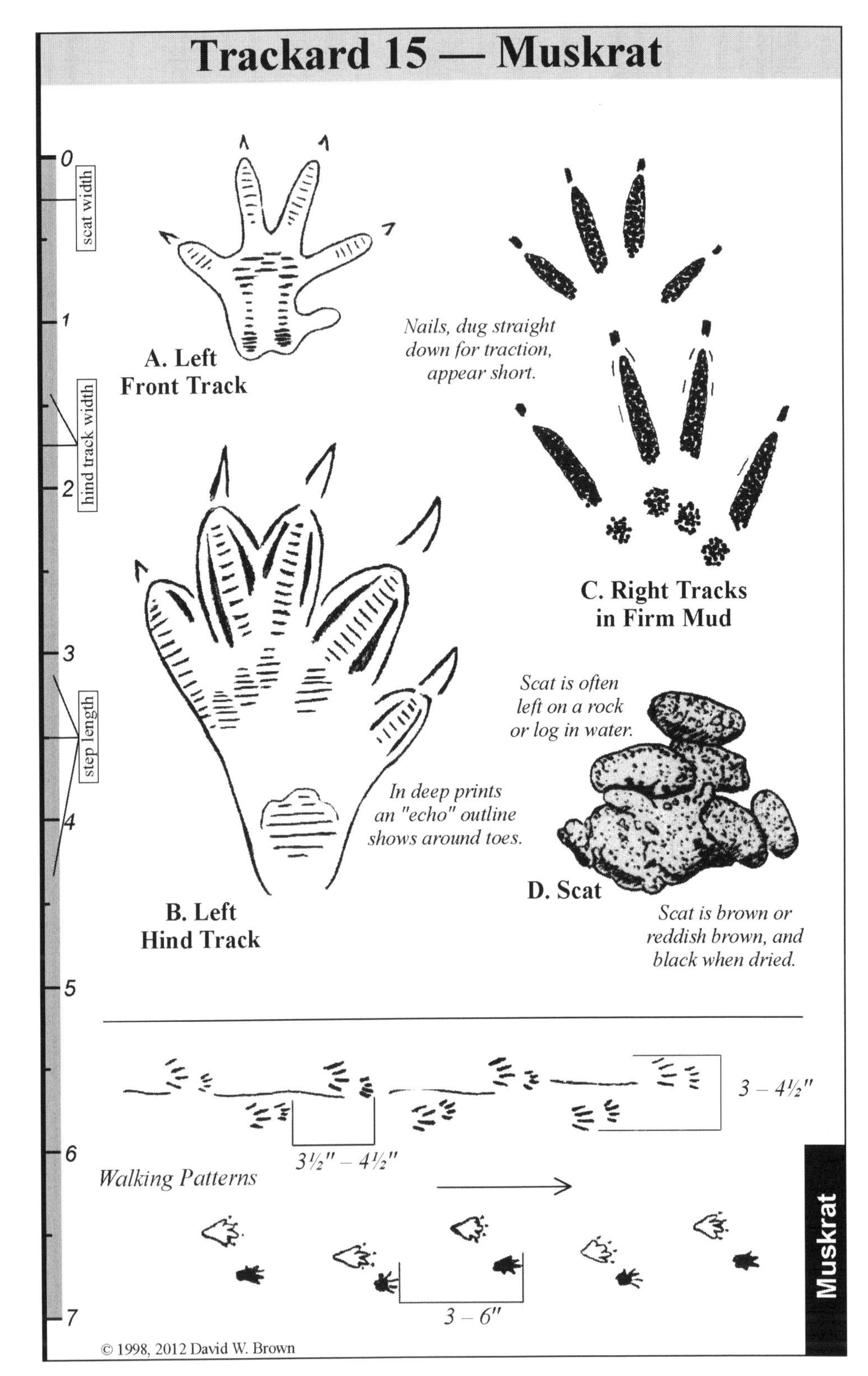

Figure 15.1. Trackard 15 – Muskrat.

Figure 15.2. Muskrat tracks and scat.

E. Various profiles of front and hind prints in mud, $^2/_5$ actual size. Saugus, Massachusetts.

F. A rather large scat deposit on a rock at the edge of a bog pond, actual size. Brownfield, Maine.

The appearance of hind prints can vary widely. Figure 15.1 B, from Saugus, Massachusetts, shows a hind print in smooth streambank mud in which the muskrat had laid its claws flat and stood flat-footed so that the entire outline and padding registered. This is an exceptionally clear print with the signature of a muskrat showing clearly; not only do the phalangials register on the centerline of each toe but these are outlined by the impression of a swim aid distinctive to muskrats. This is a flap composed of stiff hairs that shows in most prints as an "echo" outline.

Figure 15.2 E shows a number of front and hind profiles in mud. These illustrations, which were traced from casts, represent a more usual appearance, with the nails of the hind foot dug straight down, making them appear short, and the heel (tertiary area) not registering at all. In the front prints the medial toe is nearly absent; in some cases it was dug straight down into the mud, in others it was withdrawn upwards from the print. In the shallow prints shown in Figure 15.1 C, from West Newbury, Massachusetts, the action of forcing the nails down into the mud withdrew the toes slightly from the surface so that only the phalangials registered without the echo outline. This resulted in a very different appearance from the other prints illustrated. In this case the toe bones taper evenly to the nail marks unlike the clublike appearance of raccoon prints under similar conditions.

Adding to the confusion created by these very different appearances is the fact that in shallow tracks either the medial or the lateral toe may not register at all, creating an impression of a four-toed animal as in Figure 15.1 C and in two of the hind prints in Figure 15.2 E.

Trails

Muskrats usually walk. The upper trail at the bottom of Trackard 15 (Figure 15.1) shows a shallow walking trail where the hind print has understepped the front, leaving a very characteristic pattern in the mud. Viewed from the direction of travel, such a trail often looks like a series of fans receding into the distance. Note that the front straddle is narrower than the hind, a characteristic of the trails of many rodents, which tend to have wider pelvises than chests.

In the trail illustrated at the extreme bottom of the card from the Saugus River, the animal has oriented its spine at a diagonal to the direction of travel in a displaced walk/trot so that front does not interfere with the passage of the hind. In this case the front prints are all lined up on one side of the trail and the hind prints on the other, much as with the displaced trotting gait of red foxes and coyotes.

One of the features that should be looked for in trying to distinguish muskrat prints in a commotion of tracks in streamside mud is the tail mark. This continuous line through the pattern is usual in warm weather. Worm trails, also found in mud, are not as wide nor are their undulations as regular. On snow or other rough, cold surfaces muskrats may raise their tails to reduce abrasion and conductive heat loss, and so winter trails may not show this distinctive sinuous mark.

When moving across a danger area, such as an open space between the water and desired vegetation, a muskrat can use a faster gait such as a lope or bound. The resulting patterns will be compact 4X groupings with about a foot between each. Although the tail may be held above the ground in such a gait, the size difference between single-registering front and hind prints should be obvious and lead to identification.

Track and Trail Comparisons

Raccoon: Muskrats occupy the same habitats as raccoons; their tracks are often found in mud next to the water. Note that the impression of a muskrat phalangial is a smooth taper toward the end while the extended toe of a raccoon shows parallel sides, giving the impression of a blunt club. This seemingly subtle distinction becomes quite noticeable with only a little practice.

Raccoons make their living by patrolling the shoreline for crayfish, mussels and so forth. Their trails normally parallel the edge of the water, while muskrat trails, by contrast, are usually perpendicular to the shoreline as the animal leaves the water, crosses the mud strip and gathers grasses and other succulent vegetation just beyond before returning to the water. Land is a danger area for these semi-aquatic mammals and so they spend as little time on it as possible lest they fall victim to a mink or other predator patrolling the bank. If a few prints are available, the different walking patterns of the two animals will be diagnostic.

Occasionally muskrats do go ashore for an extended "ramble," visiting fields or other areas adjacent to the water and exposing themselves to extreme danger in the process. In winter such excursions may be a reaction to starvation as aquatic vegetation in the home pond is depleted, forcing the animals, especially the socially inferior ones, to forage more widely to sustain themselves.

Scat

The droppings in Figure 15.1 D are typical of muskrat: smooth brown ovals when fresh, often drying to a grainier reddish brown and eventually desiccating to small black pellets. They usually appear singly or in a small clump but Figure 15.2 F shows a large deposit from a bog in Brownfield, Maine. The distinctive placement is often on a flat rock or fallen log in the water near shore.

Dens

In marshy areas muskrats build lodges, domes of mud and cattails or other soft vegetation. The building material distinguishes them from beaver constructions since the larger species of rodent uses debarked wood rather than softer stalks and leaves that apparently are the densest things muskrats can cut with their less-powerful jaws. The material in these mounds has led to the speculation that the presence of cattails in a water body is mandatory for the tenancy of muskrats. However, muskrats will also readily dig and den in tunnels in the bank of a stream or pond. These burrows have underwater entrances with a tunnel that ascends gradually over several yards to a chamber above the water level. The entrances are often found during drought that lowers the water level, exposing more of the bank, or when the roof of the tunnel or chamber collapses underfoot. Minks may appropriate these burrows for their own purposes, perhaps killing the original occupants. Muskrats may also move into an active beaver lodge and live uneasily with the host species.

Sign

In the winter muskrats often push up dark vegetation such as lily pads through a weak place in the ice of a frozen pond. This dark spot absorbs solar radiation, keeping the ice soft enough throughout the winter for these animals to enter and exit the water.

Tubers of aquatic plants such as pickerelweed, arrowhead and cattail are dug up and removed to a secure feeding or storage location. The de-tubered stalks of these plants may remain at the digging site, and incisor marks may help distinguish this work from that of waterfowl.

Habitat

Muskrats are mostly vegetarians, feeding on tubers, grasses and leafage, either emergent (with their roots in the water) or close enough to shore that they can be harvested with minimal risk. They may also feed on slow-moving invertebrates like mussels and perhaps aquatic snails. The flat water of beaver ponds and marshes as well as slow-moving streams may host these animals. Fast moving streams tend to be avoided; they do not have aquatic plants growing along their margins, and muskrats are not as strong swimmers as beavers, finding it difficult to make headway against a fast current. Also, such streams are prone to frequent flood and ebb that leaves a sterile band along the bank, a danger zone for muskrats that would have to cross it to get at food farther up the bank. Reeds, sedges, rushes, plants with tubers and other vegetation growing in or near the water are the usual requirement for this animal, which may be present even if lodges made of mud and cattails are not found.

Trackard 16 – Gray Fox

The gray fox is a mysterious animal; because it has not been regarded as a valuable fur-bearer and because of it secretive habits, it does not seem to have been well studied. Its range extends over most of the country except the Northwest. Although present historically in northern and central New England, its numbers declined with colonial deforestation. The rematuring of the northeastern forest has led to its spread back into its ancient range in increasing numbers. It favors forested habitat because it has an unusual skill for a canid: it can climb trees. This skill provides it with a hunting advantage over red foxes, which I have occasionally seen it displace in mature woodlands.

Gray foxes are often confused with the so-called "cross fox," a morph of the slightly larger red fox that has mixed gray and red fur. The reddish ruff around the face of gray foxes increases the confusion. Whatever the color of their pelage, however, all red foxes have a white tip on their brush while grays have a black tip at the end of a black line down the length of the tail.

Tracks

Gray fox prints show four toes arranged symmetrically. The pads themselves are naked at all seasons. Neither the medial nor the lateral pads show the angularity on their inside edge that is often a canine characteristic (Figure 16.1). Instead, all pads are oval and appear easily deformable under load. A number of profiles are given in Figure 16.2 G and H in order to illustrate the range of potential shapes that may be encountered. The deformability of prints shows that gray foxes share with cats the rather thick, rubbery pads that are an adaptation for absorbing the shock of jumping down from things. The depth of these pads means that their shape will change with the depth of the print to a greater degree than with other canids. Another feline trait of the gray fox is its retractable claws, a feature that keeps them sharp for climbing. When the animal needs traction, as in the slippery mud

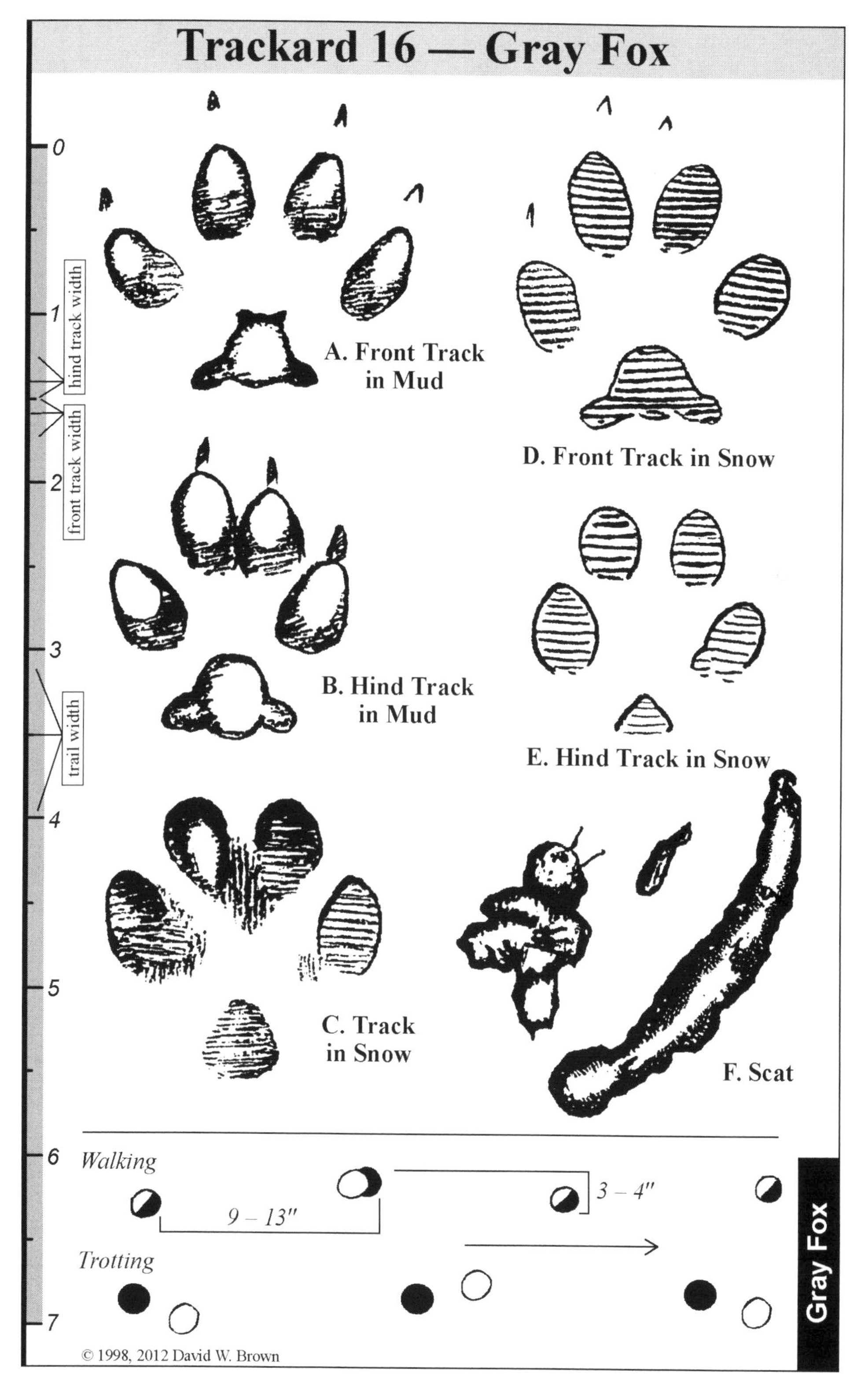

Figure 16.1. Trackard 16 – Gray Fox.

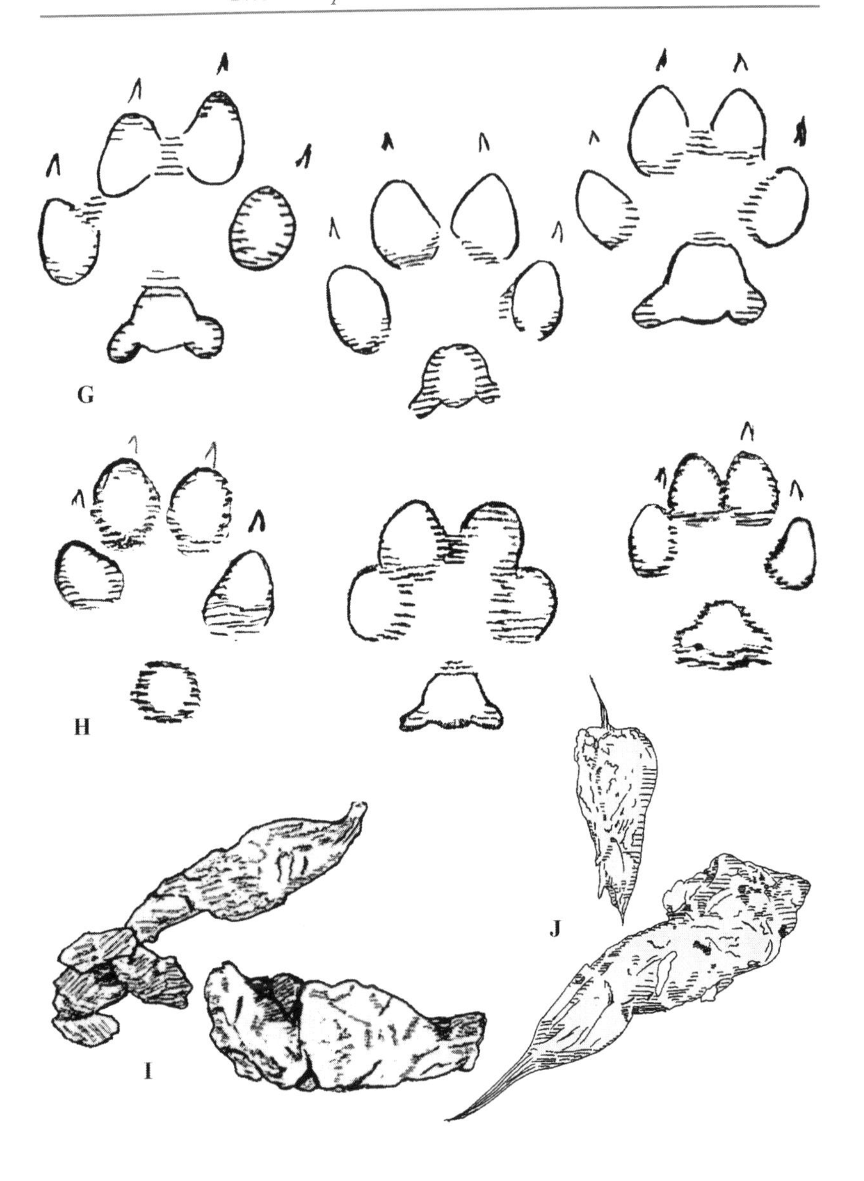

Figure 16.2. Gray fox tracks and scat, about $^4/_5$ size.

G. Various front prints in mud. Saugus, Massachusetts.

H. Various hind prints in mud. Saugus, Massachusetts.

I. Dry winter scat, tan/brown. Conway, New Hampshire.

J. Fur scat with small bone chips, Arizona.

in which Figure 16.1 A and B were recorded, they can extend these claws. At other times, however, they may not appear at all, as in Figure 16.1 C, a direct registration of hind into front, photographed in snow.

The general impression of the front print of a gray fox is of catlike roundness as opposed to the usual oval shape of red foxes and coyotes. A fringe of fur along the trailing edge of the secondary pad, however, may register in loose snow and make the print seem longer and more oval than it really is. Since the front feet bear the weight of the fox's head, the front tracks often show toe pads splayed more widely than the hind. This splaying is a good distinguishing feature for this species when it is compared to the normally tight prints of a red fox. Surface conditions exaggerated the splay in Figure 16.1 A; the more usual spread is shown in the snow track in D to the right and in Figure 16.2 G. The secondary pad is disproportionately small for a canid and is often de-emphasized in a print by the animal's tendency to toe-down as it steps, like a woman in high heels. This pad is often described as a "winged ball." The central circular area is protuberant, however, so that in shallow prints the wings may not register.

The hind feet bear only the weight of the animal's haunches and so the toes tend to spread out less. In fact, the hind print sometimes appears quite narrow compared to the front, an effect much more noticeable in the gray fox than in any of the species with which it may be confused. The hind prints shown in Figure 16.2 H, from casts collected in Saugus, Massachusetts, show this effect. The four toe pads are grouped closely together with a large space between them and the dot of the secondary pad. A combination of a round front print and a long, narrow hind is often diagnostic for a gray fox identification. Because of the light loading of the hind secondary pad, the wings of the central lobe may not appear in the print at all.

Trails

In their hunting forays gray foxes tend leave an alternating pattern of direct and indirect registrations more often than red foxes, resorting less commonly to the displaced trotting and loping of their more colorful cousins. The usual pattern is illustrated at the bottom of Trackard 16 (Figure 16.1). It is often said that grays leave a wider trail width than reds. Careful examination of their trails, however, shows this to be something of an illusion. All walking animals have some straddle to their walking gait, however narrow this may be in the case of foxes. The gray fox's smaller size results in a slightly shorter walking step length, and this compression exaggerates the zigzag appearance of the trail. One clue that is helpful is that on firm surfaces and in thin snow cover walking grays tend to be sloppier than reds, occasionally indirect and even double registering. The absence of grays from the snowy north regions of their ancestral range for a century or so may have reduced the instinct for efficient foot placement that other predators of the North Country show in the winter.

I have noticed over the years that trotting gray foxes tend to keep their body aligned with the direction of travel rather than displacing as is the case with red foxes. This is a characteristic shared with other predators that hunt in dense cover, like bobcats. The result can be a succession of direct, indirect or double registrations, but commonly showing some overstep and with a longer step length than in their walking gait. On an even, firm surface such as a frozen pond, gray foxes occasionally straddle trot as well. The resulting pattern is shown at the extreme bottom of Trackard 16. Once again, this is an aligned trot, where the hind feet pass by on either side of the planted front, rather than on only one side as with the displaced trot common among red foxes.

Track and Trail Comparisons

Red fox: Red foxes have larger prints than gray foxes, and furthermore they are oval with the long axis paralleling the direction of travel. Gray fox front prints, at least, usually appear nearly round, more like prints of a cat than other canids. The feet of red foxes are fur-covered at all seasons, showing diffuse pad outlines compared to the distinct ones of the gray. The secondary pad of a red is proportionately larger than that of the gray and usually shows a bar running across its breadth. The wings of the gray fox's secondary pad, when they register lightly, may be mistaken for this bar, but a close look should show a lack of continuity across the ball.

The normal walking step length of a gray fox ranges from 9–13 inches, compared to the 15 inches usual for reds, but in soft conditions the animal may stretch or shorten its step. In the absence of good print detail, occasional indirect registrations and often a more zigzag appearance to the trail may help to tell the difference. If the trail leads to a urine spot, identification can be confirmed with a whiff (see Scat below).

Gray foxes are less bold than reds, preferring the concealment of brush or deep forest shadow to the easier terrain of fields, frozen ponds and human trails frequented by red foxes. While reds exploit these level surfaces with a flat and efficient displaced trot, grays, like bobcats, tend to stay in rougher terrain where a bouncy aligned gait is more appropriate.

Cats: Gray fox tracks, especially when nails do not show, are easy to confuse with those of a small bobcat or even a large house cat. All three have oval toe pads and leave round prints when they walk. Cat tracks usually show a larger secondary pad relative to the size of the toe pads. However, in cases where the gray fox comes down hard on its front foot, in a high trot for instance, its spongy secondary pad may flatten out and appear larger. Furthermore, the wings of the central protuberance in vague tracks can easily be confused with the wavy posterior edge of a cat's print. Generally, however, a close look at these medial and posterior lobes, or "wings," should show them to be narrower than the more substantial lobes of cats.

A cat walking in snow can also leave a trail that is quite similar to that of a gray fox: direct and indirect registrations with the appearance of a rather wide trail width. In addition, the normal step length for a house cat of about 8 inches approaches that of a small gray fox, and when the fox stretches its step, in soft snow for instance, its length can approach that of a bobcat. Few rules (or measurements) in tracking can be applied arbitrarily, but rather must be imbedded in the circumstances: the gait of the animal, the surface in which the trail was impressed and so forth. Was there a reason for the animal to stretch its step? Was it using a gait with a high arc, like bounding, the impact of which would spread out the pads, or was it a flat, low-impact gait like walking or straddle trotting? Add this information to the total perception of the situation and make a judgment, aware that one will not always be right. As I often insist in this guide, difficult identifications in tracking are made on the sum of the evidence rather than on a single detail.

Scat

Two examples of winter scat of gray foxes are represented in figures 16.1 F and 16.2 I, both photographed in winter on the Conway

River in northern New Hampshire. There is little to distinguish them from red fox scats other than a marginally smaller size. They seem to be distributed randomly rather than being placed on prominent objects for marking as is the case with reds and other carnivores of similar size. The fur scat presented in Figure 16.2 J is from the chaparral desert of central Arizona, a region where gray foxes replace both the red foxes of the North and kit foxes of the desert farther south. In this region it is common to find fur scats mixed with dry berries and seeds.

Although gray foxes do use their urine for marking, it lacks the strong smell of red fox deposits. Rather it is so mildly scented, to the human nose at least, that it is often difficult to find. Watch for a change in walking pattern next to a raised object such as a snow-covered stump or a fir seedling. The deposit is usually small, pale yellow or even greenish yellow with an odor that can vary between mild musk and chemically treated paperboard; this latter is the same as fisher urine (and unfortunately some bobcat urine, as well). When male gray foxes are looking for mates at mid-winter and advertising with their urine, the mark may be substantially larger and more conspicuous. The odor, too, may be more acrid, to my nose something like the smell of sweaty sneakers.

Habitat

Because gray foxes can climb trees, they enjoy a hunting advantage in woodlands over the competition: they can leave the ground in pursuit of arboreal prey such as squirrels. At times when rabbit or hare populations go into one of their periodic dives, the gray can outcompete the red fox with this talent and may displace it over a forested range. Look for gray fox tracks in bottomlands and brushy clearcuts where small prey abound as well as in upland forests where its climbing ability can be put to advantage. In the winter, south-facing hillsides tipped up to the sun are attractive to grays for their faster melt-back and possibilities for sunning.

Trackard 17 – Mink

Minks inhabit almost all of the United States and Canada except the Southwest and the extreme Southeast. Like other members of the weasel family, they have short legs and long bodies adapted to fitting into holes for hunting and denning as well as entering and exiting water through small openings. Smaller than a house cat, these nervous little animals have lustrous dark brown fur for which they are justly famous. A small patch of white under the chin relieves the otherwise dark appearance. Although minks can swim, they are less dependent on fish and other aquatic prey than are their larger cousins, the otters. They display the weasel's ferocity in tackling prey as big as muskrats and ducks. When distance obscures relative size, a mink swimming across a pond can be told from an otter by the fact that the mink will paddle along in a flat plane with tail stretched out behind and with much rippling due to the rapid turnover of their relatively small feet. Otters, which are much larger and have big webbed feet, cruise sedately or undulate like sea serpents as they work a pond.

Tracks

Like other mustelids, minks have five toes on each foot with the medial fifth toe de-emphasized and retarded. Often this toe either registers as only a dot or not at all. When it shows, the track appears asymmetrical. Claws are short, sometimes joining the toe in a teardrop shape common to other weasels, as in the hind print shown in Trackard 17 (Figure 17.1 B), recorded in Brownfield, Maine. The secondary pad is irregular, emphasized on the lateral side, de-emphasized and retarded on the medial.

The front foot, bearing the weight of the head, typically splays its toes more than the hind. This is clear from the front prints shown in Figure 17.1 A and B, both sets from the Brownfield Bog. When analyzing a track pattern for gait and speed, it is often necessary to distinguish front from hind. Bisecting each of the two foremost

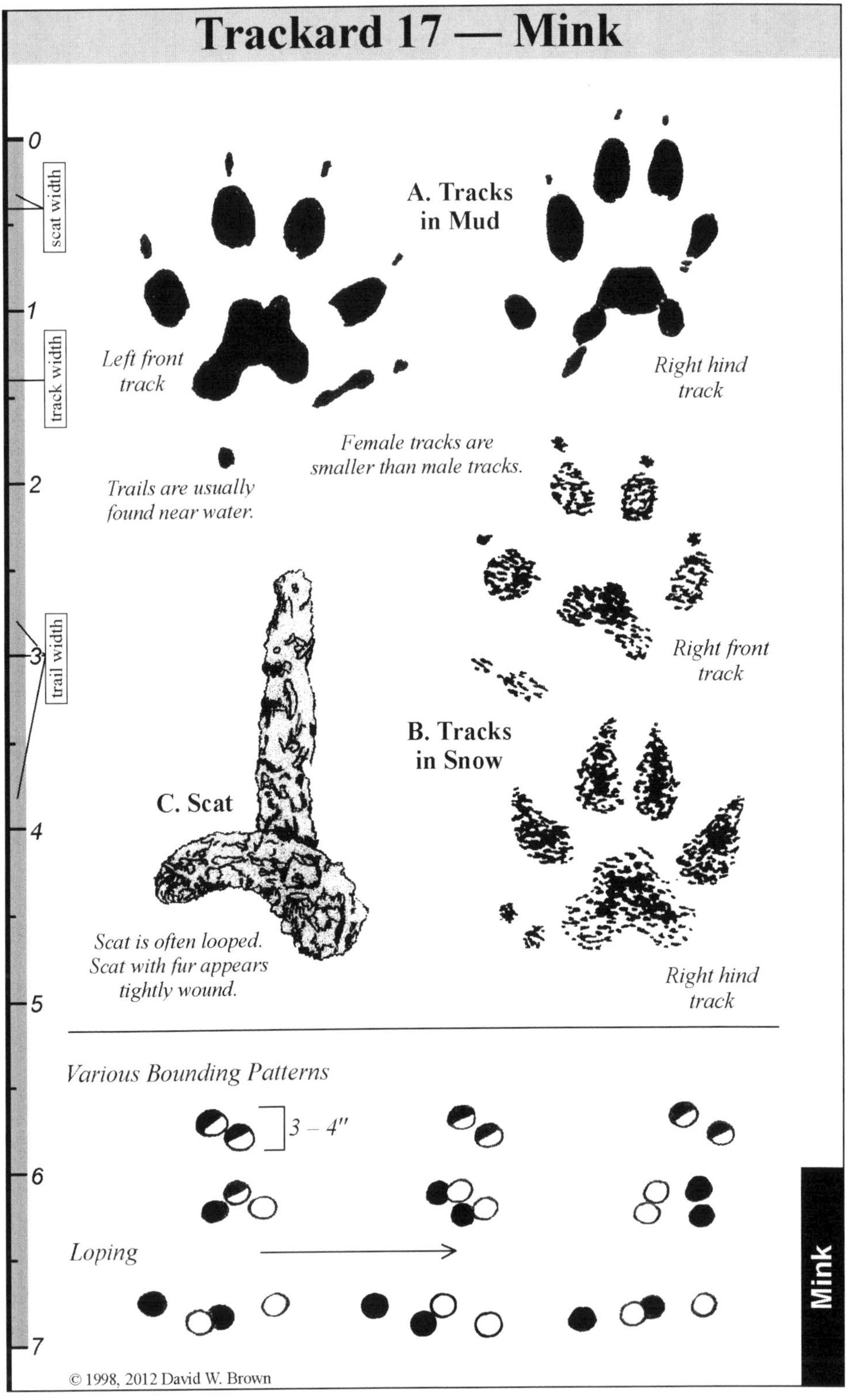

Trackard 17 — Mink
0
1
2
3
4
5
6
7
scat width
track width
trail width
A. Tracks in Mud
Left front track
Right hind track
Female tracks are smaller than male tracks.
Trails are usually found near water.
Right front track
B. Tracks in Snow
C. Scat
Right hind track
Scat is often looped. Scat with fur appears tightly wound.
Various Bounding Patterns
3 – 4″
Loping
Mink
© 1998, 2012 David W. Brown

Figure 17.1. Trackard 17 – Mink.

central toe pads along its long axis will usually show parallel lines for the hind and splayed lines for the front. Also the secondary pad of the front print commonly appears more poorly developed on the medial side than is the case with the hind print. Sometimes the front print also registers a tertiary pad on the lateral side as the small dot shown at the bottom in Figure 17.1 A, but this is not reliable.

Minks are sexually dimorphic, a typical adult male being about a third larger than the female. Front track width for the male, whether measured perpendicular to the direction of travel or across five toes, is often 1³/₈ inches wide, a measurement that shows up again and again for this species. All the prints on Trackard 17 are those of adult males; the front print of a female or independent juvenile may measure as much as ¼ inch narrower.

Trails

In non-snow seasons when minks are traveling shorelines littered with debris such as boulders and fallen limbs, their gaits will vary often in response to these irregularities. In winter, however, when snow smoothes the terrain for these short-legged animals and greater gait efficiency is critical to survival, mink patterns become much more regular. In "deep" snow, which for a mink means a track depth of only a couple of inches, the animal will resort to the methodical bounding pattern shown in the uppermost trail at the bottom of the card. The resulting pattern of two direct registrations on a repeating slant is shown in Figure 17.2 G , recorded in snow at Petersham, Massachusetts. In this gait the front feet land on a diagonal, packing down the snow for the hind feet, which follow them into the same outline. The twice-packed snow then serves as a firm platform from which the mink can launch itself into the next bound.

On firm surfaces such as crust or a skim of snow over ice, a mink may use a loping gait, the lower profile of which is much more efficient than bounding. Various versions of this gait pattern are shown on the bottom line in the Trails section of the card (Figure 17.1).

Mink trails most often parallel the water's edge, sometimes cutting across peninsulas encountered along the way, a habit minks share with otters. Their trails may also wind through cattail marshes or other wet habitats. Minks visit a circuit of wetlands in their range, so that their trails may be found well away from water as they connect up these areas. In early spring when males are searching out the natal

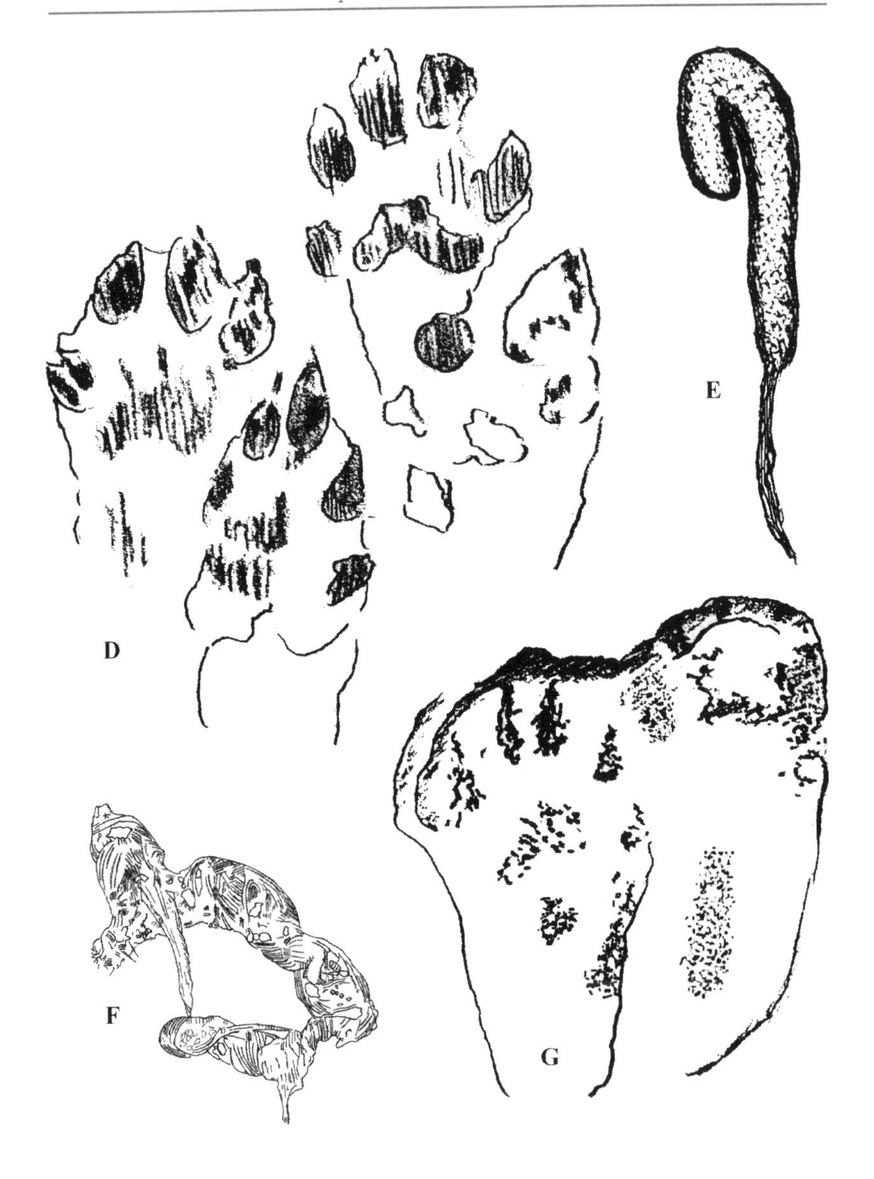

Figure 17.2. Mink patterns and scat, about 2/3 actual size.

D. Four-print understep bound in thin snow over ice. Petersham, Massachusetts.

E. Composite scat. Lincoln, Massachusetts.

F. Tightly wound fur scat. Cambridge, Massachusetts.

G. Direct registering bound in snow, depth 2–3 inches. Petersham, Massachusetts.

dens of females, their trails may also be found wandering away from water.

Track and Trail Comparisons

Gray squirrel: If the vestigial medial fifth toe of a small mink does not register, its track may be mistaken for the front print of a gray squirrel with its toes contracted longitudinally so that the characteristic toe bones don't register. Both will show four toes and a more or less M-shaped secondary pad. Careful search for the fifth toe mark often solves the puzzle, as does inspection of the immediate area for the hind print of a squirrel, which has a group of three advanced central toe pads to the mink's two. Another tip-off for squirrel tracks is their relatively long nail marks.

When the track depth is shallow, about an inch or less, the hind feet of a bounding mink, not needing pre-packing for their placement, may overstep or understep the front. Some resulting patterns are shown in the middle line at the bottom of Trackard 17 and an understep pattern of this sort is presented in Figure 17.2 D. When a small mink understeps in a "lazy bound," as in the right pattern at the middle bottom of Trackard 17, care must be used to distinguish the four-print pattern from gray squirrels, which show a similar pattern. For reasons that go beyond the scope of this guide, minks sometimes square off their front prints in an understep so that they are arranged more or less perpendicular to the direction of travel, increasing their similarity to the familiar four-print pattern of squirrels. The confusion may be compounded by the relative hind-front print sizes, that is, the retarded front prints of the squirrel can look like the smaller hind feet of the mink, which appear in almost the same position in the pattern. The larger hind prints of the squirrel also appear in the same advanced position as the mink's front prints, which are larger than its hind. If track detail is not evident, compare the straddles of the two species. There should be a greater difference between front and hind straddle for the squirrel than for the mink, that is, the four-print squirrel pattern will look more like an inverted trapezoid while the mink's will be closer to a rectangle.

Weasel: Although long-tailed weasels as a species are smaller than minks, the size of their tracks can approach one another since both species are sexually dimorphic. During periods when the predominant prey species is a larger animal, snowshoe hare for instance, weasel reproduction selects quite rapidly for larger size. During such times a big male long-tailed weasel may approach the bulk of a small female or juvenile mink. Since both are mustelids with similar feet, some confusion can occur. Minks are certainly more aquatic than weasels, and so habitat may be a clue. However, minks occasionally traverse the weasel's uplands and long-tailed weasels sometimes hunt wetlands. In such situations there is little to distinguish similar sized members of the two species. However, if a trail enters water or a bank hole leading to water, then identification as mink is much more likely, at least in the winter.

Weasels use all of the patterns in snow that a mink does, and a male long-tail's patterns, as well as its prints, may show measurements approaching those of a small mink. A trail of appropriate size that stays right along the edge of a pond and occasionally enters the water is certainly that of a mink. Also mink trails in snow tend to be more methodical than those of weasels, with more or less constant inter-pattern distances as well as more consistent direction.

Most members of the weasel family slide in snow, although otters use this behavior most frequently. Distinguishing the troughs of one from another is a matter of measuring their width. Otters leave 7–9-inch troughs, while mink slides will be only about 4 inches across and weasels about 3 inches. There is considerable overlap among the smaller species, but any of their snow slides that end in water can safely be assigned to mink.

Scat

Mink scat varies with the prey consumed. A trait with which to distinguish it from the scat of similar sized carnivores is its tendency to be looped or at least curved, a function of the proximity of the anus of a short-legged animal to the ground. Typical hair scat is dark and twisted into a distinctive tightly-wound appearance. Such carnivore scat will be composed of fur of small wetland prey such as bog lemmings or red-backed voles on up in size to muskrats. If bones appear on the surface they will have been finely chewed into small chips. Normal maximum diameter is around 3/8 inch. The scat illustrated in Figure 17.2 F was found on the raised bank of a small tributary to a stream. Mink regularly patrol such banks, perhaps as a way of getting maximum benefit from their hunting time. Such locations are on the edge of contiguous habitats where a greater variety of prey may be found, including small mammals on the shore and fish or crayfish in the water.

Scat from feeding on fish or crayfish will show scales, fish bones or reddish chitin. Since fish or crayfish are easier to catch than furry prey, scat of this sort is likely to be found much more often. An example is provided in Figure 17.1 C, from a pond edge in Concord, Massachusetts. Otter and raccoon scat on the same diet will be both larger and more copious.

Mink droppings are often found on elevated ground in or next to a wetland such as on top of a streamside boulder, on top of a beaver or muskrat lodge, on a bog berm or on the root hump of a fallen tree. The scat shown in Figure 17.1 C was found on such a root hump in water a few feet from shore. Scat may also be found near a wet hole in a bank where the animal emerged from under the ice. Shortcuts across peninsulas are also good places to look for scat. Finally, mink seem to prefer to feed in concealed situations. I have found feeding sign including scat in locations such as the trunk of a hollow tree near water and at the entrance to a muskrat burrow that has been left above water by drought.

Sign

Minks enter and leave water at all seasons. In the winter when ponds are frozen over, a mink will gain access through bank holes or weaknesses in the ice such as often occur on shores with southern exposures and overhanging evergreens that absorb and reradiate solar warmth. Minks know all the bank holes in their range. Scat, partially eaten fish or tadpoles and tracks can sometimes be found at their entrances, which are usually about 4 inches in diameter. Minks also enter water through fractures in the ice over a beaver pond caused by settling as water seeps out through the dam. They may even use the auger holes of ice fishermen while feeding on any fish guts or bait left behind.

Dens

Minks den both for birthing as well as for resting during the daytime. Birthing dens are often in abandoned or appropriated bank burrows of muskrats and beavers. Small enough to be regarded as prey by hawks and owls, minks may lay up during the day in any cavity that provides them concealment and protection. Such resting dens may be in the hollow bole of a tree or even in a crevice in an old stonewall. Both sorts of dens, for resting and birthing, are invariably near water.

Habitat

Minks are usually associated with and to a great extent dependent upon water even though they may occasionally be found traversing or even occasionally hunting nearby uplands. Healthy lakes, ponds, streams and marshes usually support a small population. The resurgence of beavers in the East has resulted in much ideal habitat for minks, whose recovering numbers seem to be tracking those of their host. Beaver ponds provide many kinds of food preferred by minks such as fish, crayfish, frogs, tadpoles, bog lemmings and other voles, muskrats, ducks and snakes.

Trackard 18 – Cottontail Rabbit

There are nine species of native rabbits in the United States, all members of the same genus, ranging from the increasingly rare New England cottontail to the desert cottontail of the Southwest and the marsh rabbit of the Southeast. All have similar feet, so that distinguishing among their tracks and sign is a matter of considering size, range and habitat. The Eastern cottontail is presented here as representative of the genus.

This species, which got its name from its white puffball tail, is common in brushy edges that it uses for cover. There in winter it survives by gnawing bark and nipping twigs while in spring and summer it can make forays from its hiding place out into meadows in search of succulent growth. Actually, in the Northeast there are two species, the New England cottontail and the introduced Eastern cottontail, the former now scarce in the region. The differences between the two are of interest mainly to taxonomists, but include smaller, heat-conserving ears in the New England variety, a function of its more northerly distribution. The northern part of the cottontail's range coincides with that of the snowshoe hare. Rabbits differ from hares in bearing their young altricially, that is, blind, naked, helpless and totally dependent on their mother. Hare young are born fully furred, eyes open and ready to run. Collectively both species of rabbits as well as the hare are grouped under the heading of "lagomorphs."

Tracks

Rabbit feet are covered with dense fur composed of long stiff hairs that point down and forward. This covering largely conceals the toe pads, which tend to register as four vague protuberances with a retarded fifth on the medial side sometimes showing as well. Figure 18.1 A, photographed and cast in streambank mud at Saugus, Massachusetts, shows these characteristics. Each nail registers as a point at

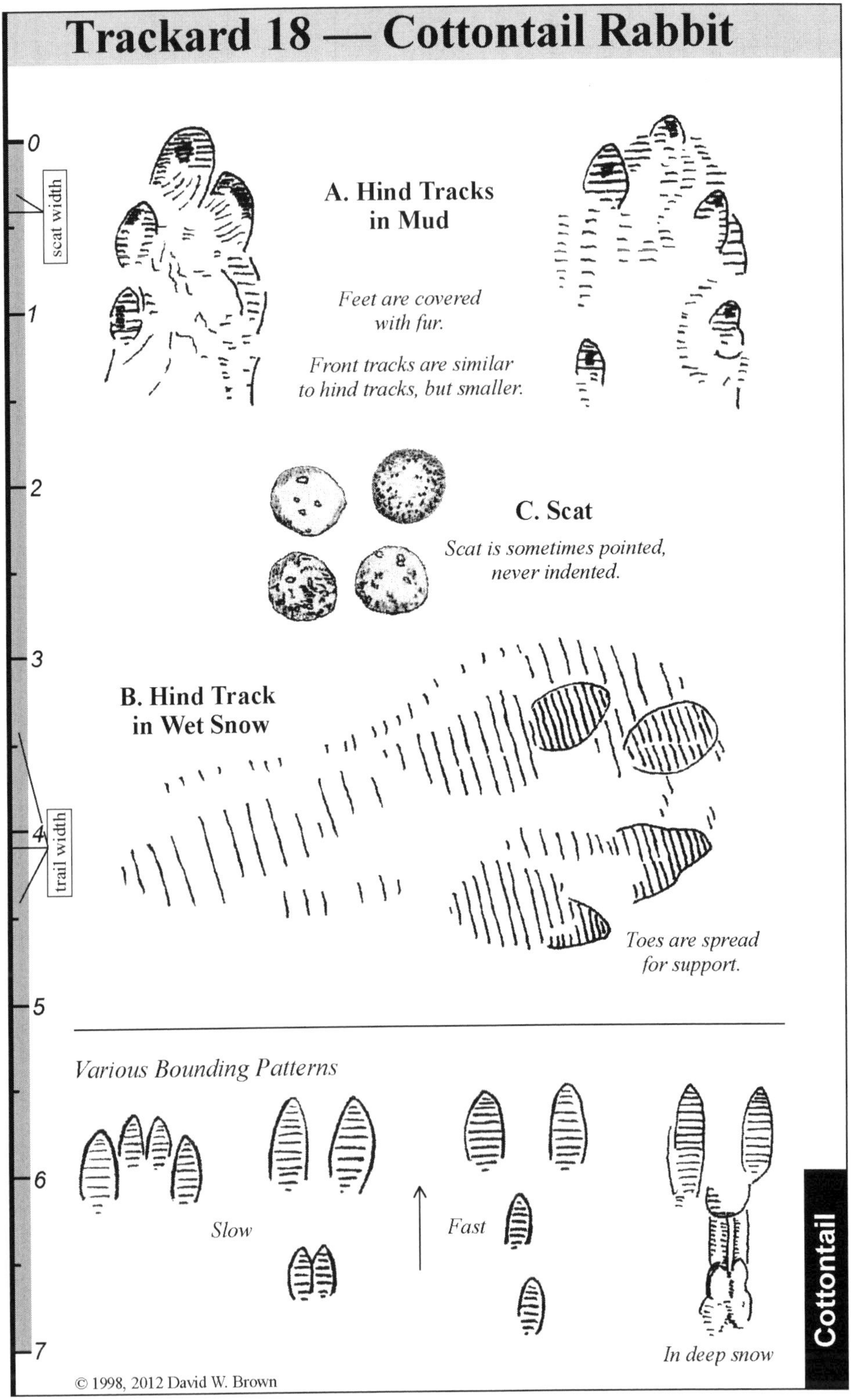
Trackard 18 — Cottontail Rabbit
0
1
2
3
4
5
6
7
scat width
trail width
A. Hind Tracks in Mud
Feet are covered with fur.
Front tracks are similar to hind tracks, but smaller.
C. Scat
Scat is sometimes pointed, never indented.
B. Hind Track in Wet Snow
Toes are spread for support.
Various Bounding Patterns
Slow
Fast
In deep snow
Cottontail
© 1998, 2012 David W. Brown

Figure 18.1. Trackard 18 – Cottontail Rabbit.

the leading edge of the associated pad and is indistinguishable from it. The four nails are arranged asymmetrically with a pronounced bias to the lateral side, particularly when the toes are closed. When they are spread, for support on collapsible snow for instance, as in Figure 18.1 B from West Newbury, Massachusetts, the arrangement appears more symmetrical. On firmer snow, cottontails will keep their toes closed for heat conservation, and the resulting prints often show a point at the front end, making them appear spear-shaped, the result of the long foot-fur tapering to a point like an artist's brush. This effect is shown in some of the patterns at the bottom of Trackard 18 (Figure 18.1), recorded in thin snow over ice at Lincoln, Massachusetts. Secondary and tertiary padding is diffuse, registering merely as a depressed area.

Figure 18.2 D and E show a typical bounding pattern, the two front prints of which were recorded at different times in snow at Lincoln. The left front print shows pads spread for support while, for comparison, the right shows closed pads. A rabbit's front feet are smaller than its hind, a fact that may be disguised on occasions when the front toes are spread while the hind are not. Front prints generally measure between $^{5}/_{8}$ inch to 1 inch wide.

Trails

Even more so than squirrels, lagomorphs use a bounding gait almost exclusively. The trail section at the bottom of Trackard 18 shows several 4X patterns that are compressed fore and aft at slow speed and spread at high speed. In distinguishing rabbits from squirrels, the arrangement of the front feet at the rear of the pattern is significant. Either these prints register side-by-side and nearly touching, yielding the general shape of a broad V, or one is almost directly in front of the other, giving the pattern a T-shape.

This effect is the result of the very narrow chest of rabbits and hares. In the bounding pattern shown in Figure 18.2, the hind straddle is a typical 4 inches. (Note that the illustrated pattern is a composite of prints from different places and conditions.)

The rightmost illustration in the trails section of Trackard 18 shows a bounding pattern in deep, soft snow. Note the rounded tail impression just behind the hind prints and the troughs forward of the front prints where the front feet dragged through the snow as they passed forward into the next bound.

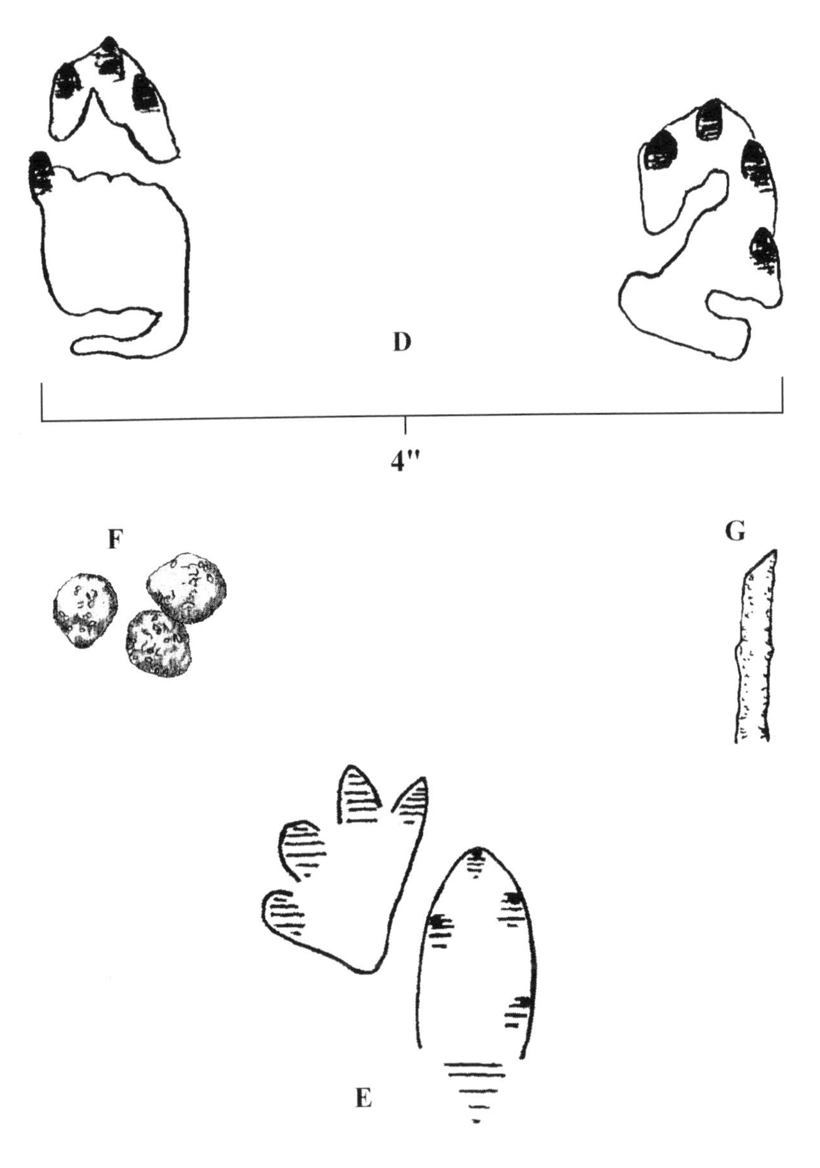

Figure 18.2. Cottontail bound pattern and sign, about 7/8 actual size.

D. Hind prints in mud, from cast. Saugus, Massachusetts.

E. Front prints in snow, from field notes. In left print toes were spread for support; in right, toes were closed. Lincoln, Massachusetts.

F. Scat with pointed ends, in New England cottontail habitat. Mount Wachusett, Westminster, Massachusetts.

G. Typical rabbit/hare twig cut. Rodents make similar cuts.

Track and Trail Comparisons

Gray squirrel: Squirrels usually show distinct pads, long nails well ahead of the toe pads, and often phalangials between the primary and secondary pads. Rabbits show none of these. However, gray squirrel hind prints are about the same size as those of cottontails, about 1 inch across, and the two species' trails often cross, showing similar bounding patterns. In vague tracks, melted out in old snow or diffuse in dry fresh snow, the two species can be hard to tell apart. In such a trail, note that the leading edge of a squirrel's hind print is always blunt, the result of the even arrangement of the central toes compared to the more rounded or spear-shaped leading edge of a rabbit's closed-toe print.

Gray squirrel and cottontail bounding patterns are superficially similar, both having a hind straddle of about 4 inches. However, squirrels have broader chests than rabbits. When their front prints register side-by-side, there is usually, but not always, a space between, giving the pattern a W-shape more than the broad V- or T-shape typical of rabbits. Furthermore, squirrel front feet often register on a distinct diagonal that increases with speed while maintaining lateral space between. Squirrels also have very flexible hind feet, typical of tree-climbing animals. This translates on the ground into a tendency for the hind prints to splay outward, particularly in shorter, slow speed patterns. Rabbit feet are laterally rigid, built for straight ahead speed, and will leave hind prints that are usually parallel. No rule works all the time, however; occasionally rabbits, foraging slowly over the ground, will also splay their hind feet.

Snowshoe hare: Although both species are lagomorphs and leave similar shapes and patterns of prints, snowshoe hare tracks are much larger, with a minimum close-toed hind print at least ½ inch wider than that of a maximum spread-toed cottontail. This can often be gauged quickly by extrapolating from one's thumbprint. Mine is very close to the size and nearly the shape of a cottontail's foot.

Scat

Cottontails defecate where they feed. If they have located good browse in one spot, quite a collection of scat may accumulate. Individual droppings will be round or slightly flattened spheres such as those shown in Figure 18.1 C from Lincoln. Although occasionally scat is found with a pointed end on each sphere, rabbit droppings never show indentations. A collection of this sort of pointed scat is shown in the Figure 18.2 F, photographed in habitat typical for the New England cottontail. Most rabbit scat has the consistency of compressed yellow or brown sawdust.

Scat Comparisons

Snowshoe hare: Snowshoe hare scat usually measures larger than cottontail scat: ½ inch to the cottontail's usual 3/8 inch and, in my experience, is normally more flattened. However, in very good habitat cottontails may leave larger and sometimes quite flattened droppings, so identification solely on the basis of measurement and appearance is chancy. As with most identifications in tracking, look at the sum of the evidence including habitat. I find snowshoe hare sign, but seldom cottontail, in dense evergreen cover such as laurel thickets and regenerating conifers. It also is common in forest openings with blueberry patches. On the other hand, cottontail sign is more common in deciduous thickets where the rabbit's brown streaked winter pelt blends better with the background. In summer, rabbit trails, but seldom those of snowshoe hare will be found in open fields and meadows.

Deer: Although deer pellets usually are deposited in quantity at a given spot, a lagomorph feeding at a productive site may leave a large number of droppings as well. Close inspection, however, will show that at least some of the deer droppings are not only pointed, but also cylindrical or indented at an end or side. I have never seen either of these features in the droppings of cottontails or snowshoe hares.

Feeding Sign

When lagomorphs and rodents nip twigs, the results look like the work illustrated in Figure 18.2 G. In order to manage a long twig, the animal is forced to hold it diagonally in its mouth so that it can make one or more diagonal cuts with its sharp central incisors. Holding the twig lengthwise in its mouth would only result in mashing it, while holding it perpendicular is impossible because the animal's head would get in the way. The result of this position is a neat diagonal cut, as in the illustration, without the square-cut and fraying typical of deer browse. Lacking upper incisors, deer must rip twigs with a jerk of the head and the results look like it. If what appears to be neat lagomorph work is high off the ground, at a level consistent with deer browsing, consider the possibility that a rabbit or hare was standing on top of deep snow as it fed.

When rabbits debark branches, the cutting is quite ragged and deep, with some of the wood underlying the nutritious cambium being gnawed as well. Porcupines, on the other hand, do the same job more neatly, just removing the edible surface. The deeper cutting into the woody underlayers by rabbits may be a function of their dentition. Unlike rodents, lagomorphs have a double set of central incisors. The outer two compare with a rodent's, but just behind them is a second, shorter set. I suspect their involvement in the gnawing process results in the deeper and more ragged cuts.

Habitat

Eastern cottontails generally inhabit dense brush, venturing into the open only if the enticement, such as green grass and clover in the spring or vegetables in Mr. Macgregor's garden in summer, is overwhelming. Conveniently, such brushy areas are often on the sunny edge of a pasture or farm fields. In this brush the animal hides by holding still and allowing its tawny-brown, mottled fur to blend with the surrounding branches and trunks. This habitat provides it with food in the form of reachable buds and tender bark as well as cover and concealment from avian and terrestrial predators. Nonetheless, it must breed prolifically to keep up with the level of predation, as this species is a staple of foxes, coyotes, bobcats, fishers and weasels as well as large hawks and owls.

Since the introduction of the Eastern cottontail from the mid-Atlantic area into the range of the native New England cottontail, the population of the latter has contracted severely, with current range maps showing it limited to widely separated "islands" of southern New England. However, I have found droppings high in the stunted boreal forests of northern New England mountains that appear to belong to this species. This leads me to suspect that there may be an undiscovered population in this habitat where wildlife surveyors have not thought to look. Certainly, such areas are too harsh for the Eastern variety and so by withdrawing to the high mountains the New England cottontail may have escaped competition with it.

Trackard 19 – Striped Skunk

Several species of skunks inhabit the United States, ranging in size from the small spotted skunk of the South and Midwest to the large hog-nosed skunk of the Southwest. The most widely distributed is the striped skunk, which is intermediate in size between the two extremes just named and can be found coast to coast. All skunks have similar feet, so that distinguishing among them is largely a matter of size and range.

Despite their rotund shape, skunks are related to weasels, a family composed mostly of quick, lithe, long-bodied animals. All have musk glands near their anus, but in skunks these have evolved into such a potent weapon of discharge that the animal no longer needs quickness for protection. Like the porcupine, which has an effective defense of its own in its quills, the skunk has expanded its girth and become slow-footed. Only if one looks just at the head and ignores the rest of the body can a resemblance to weasels still be seen. By habit the skunk is primarily nocturnal and solitary although several may be seen together at a concentrated food source. Aware of the effectiveness of its defense, it can be quite bold around humans and their pets. The spray can be delivered effectively within about 15 feet and while the skunk is facing its target.

Tracks

Figure 19.1 A and B show various prints in snow and mud. Both front and hind measure a little over an inch across for an adult. Five toes normally register, with the larger "big toe" located on the lateral side and medial fifth toe smaller than the rest and retarded. The print looks like that of a miniature bear in many respects: number of toes, their distribution, relative nail length, appearance of secondary and tertiary pads and overall outline. Both are bulky mammals that feed by digging in the leaf litter, and neither needs to move very quickly, so they have evolved similar feet as well as a plantigrade walk that can bring the whole undersurface of the foot into contact with the ground.

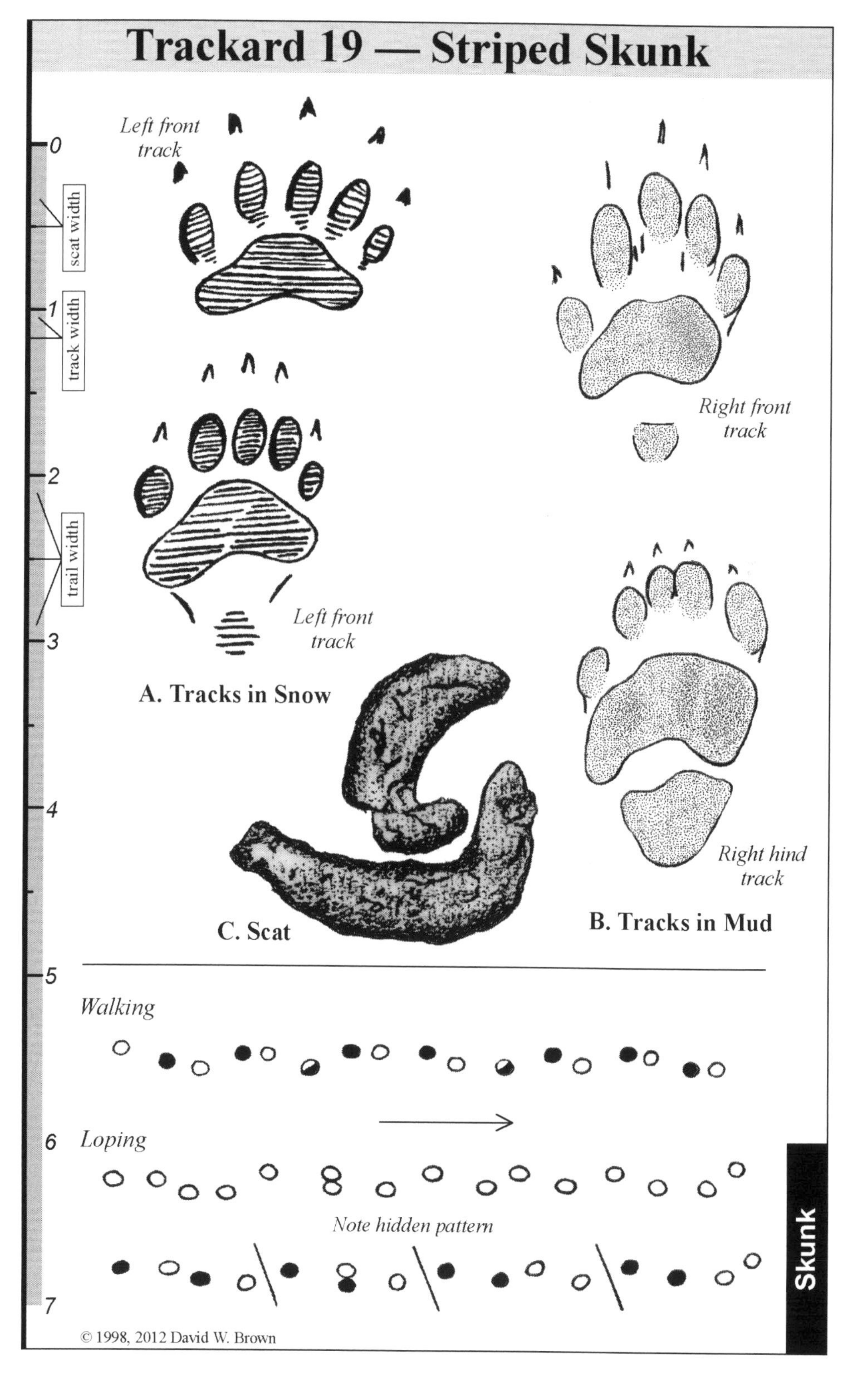

Figure 19.1. Trackard 19 – Striped Skunk.

The front feet have much longer nails than the hind since they are used for digging. The grouping of the associated toes is significant. The "big toe" on the lateral side is retarded relative to the three central toes and balances the smaller fifth toe, which is also retarded on the medial side. This makes the front print appear somewhat more symmetrical than the hind. The secondary pad is fused into one entity that will only show creases in very clear prints. It spans the width of the print and is roughly kidney-shaped although it deforms easily under load. It is emphasized slightly more on the lateral side; that is, analogous to the human foot, the "ball" is on the outside rather than the inside. Although skunks are plantigrade on both front and hind, the tertiary pad on the front foot is usually somewhat withdrawn, that is, recessed above the plane of the secondary pad and toes. When it registers, it normally does so merely as a round impression.

Hind prints are more decisively plantigrade, with the tertiary area registering in most prints even though it, too, is slightly recessed. The seam between it and the secondary pad is pronounced, appearing as a gap. The grouping of toes in the hind print is similar to the front. But, because the hind foot does not bear the weight of the animal's head and because the weight of its haunches is distributed over more pad area, the toes are apt to splay less, with the grouping of the three central toes more pronounced. The lateral "big toe" pad may be set off by a space and may appear slightly larger than the rest. The medial toe impression is much smaller than the others and is retarded more than the big toe on the other side, giving the whole track a more lopsided appearance than the front print, with a definite bias to the lateral side. Finally, the nails on the hind are much shorter than on the front. Figure 19.2 D and E show a number of profiles of front and hind mud prints from photographs and casts. The differences in the shape of the secondary pads are a result, as always, of the deformability of this fleshy pad under load as well as the variability of surfaces.

Trails

Skunks' normal food is fairly stationary: colonial insects, bugs of the leaf litter and lawn, eggs of ground nesters and so forth. And for obvious reasons these rotund relatives of the weasel do not need speed for escape. Usually, then, as they rummage around, skunks walk, but the trail they leave behind is unusual for its irregularity. Since they remain torpid for most of the winter, they have not had to learn

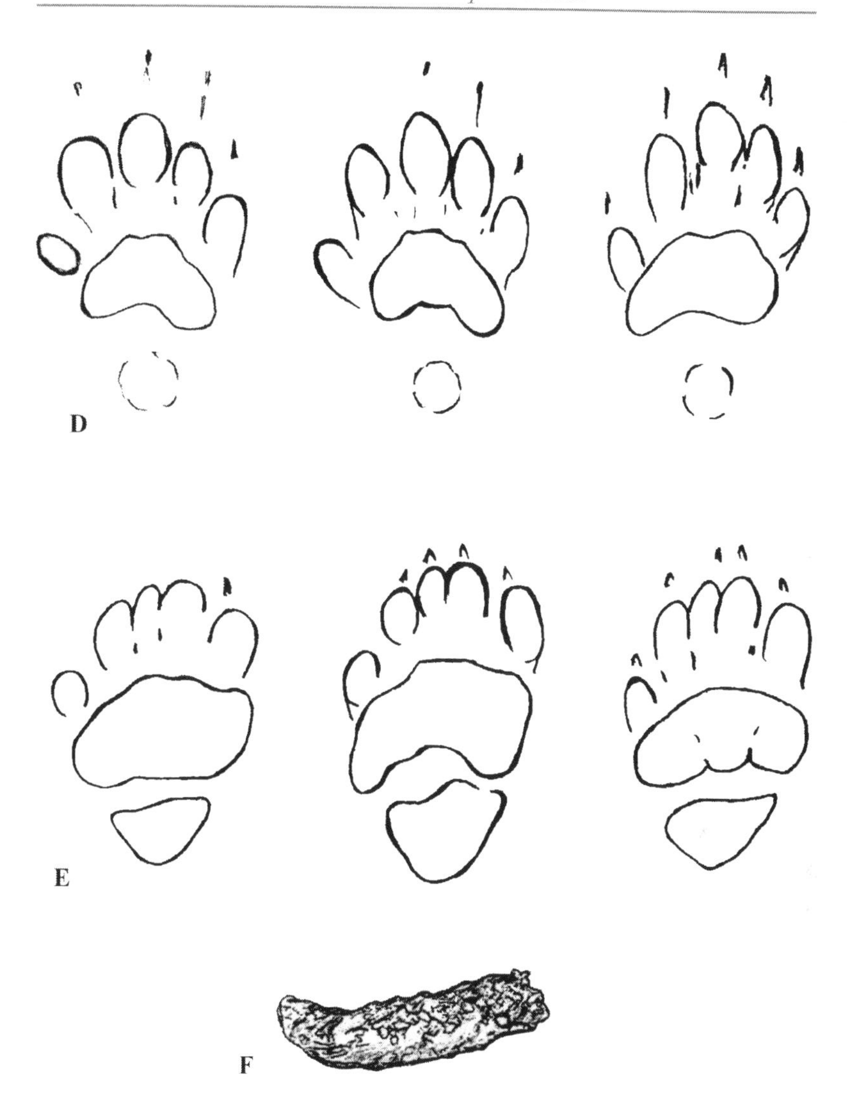

Figure 19.2. Striped skunk tracks and scat, about 2/3 actual size.
D. Front profiles in mud, from casts. Petersham, Massachusetts.
E. Hind profiles also in mud, from casts. Petersham, Massachusetts.
F. Black scat composed of indigestible ant parts. Newbury, Massachusetts.

the efficiency of direct registration in deep snow, so their tracks appear as a mix of direct, indirect, double and single registrations without any distinct pattern. This effect is reinforced by the narrowness of the trail width. When a skunk walks, its wide body teeters above its narrow undercarriage much like an overloaded third-world bus. It seems constantly to be catching itself with a quick step adjustment just before it topples over. The example of a walking trail in the bottom section of Trackard 19 (Figure 19.1), from Petersham, Massachusetts, in shallow spring snow, is typical.

When they are in a danger area or are intrigued by a distant odor, for instance, skunks can also lope. Some typical patterns in shallow snow from Saugus, Massachusetts, are illustrated at the bottom of the card, with increasing speed from left to right. The upper series of lopes is shown without identifying front and hind and without pattern boundaries. Because of the skunk's short legs and slow speed, inter-pattern spacing, which normally defines the groups of four prints in the trails of faster mammals, often fails to appear in a skunk's trail. Instead the patterns run together and look superficially similar to the walking pattern shown at the top. In the trail at the extreme bottom of the card, which is identical to the one just above, the front prints have been darkened and the pattern boundaries supplied, revealing a series of slow-loping patterns, both transverse and rotary in their order of foot placement. In attempting accurately to recreate the event from the tracks, the key to telling the difference between walking and loping patterns of a skunk is to look for the long nails of the front prints and the shorter nails of the hind. Marking a few front and hind tracks in the snow, using an "f" or "h" as appropriate, will soon reveal whether the patterns are typical 4X (four-print) lopes, such as the ones pictured, or are the more random distribution of a walk.

Scat

Skunk scats vary with diet. The scat illustrated in Figure 19.1 C was photographed at the entrance to a den that had been dug originally by a woodchuck, then appropriated and subsequently abandoned by a red fox, only to be occupied as a winter den by a skunk. It is a little larger than average for this species.

The most common food of skunks is colonial insects, and the resulting scat is a compact mass of the indigestible exoskeletons of these creatures. Such scats are typically tube-shaped with rounded or broadly pointed ends. The scat shown in Figure 19.2 F, found on an unused dirt road in July, was composed of ant parts. When such scat dries, it becomes as delicate as a cigarette ash and can easily be fractionated with a pencil point.

Scats don't seem to be used for marking and so can be found randomly anywhere the animal is feeding or denning.

Pileated woodpeckers, crow-sized birds of mature forests, commonly feed on carpenter ants that chew galleries for their nests in the heartwood of old trees and stumps. The resulting dropping also will be composed of ant chitin and be about the same size as a small skunk scat. However, along with fecal matter, birds also eliminate urea, which usually shows as white or yellow at one end of a dropping. Snake feces often show this effect, as well.

Feeding Sign

Skunks make a lot of shallow digs as they search the forest floor or a grassy field for edibles such as bugs or grubs in the earth or leaf litter. Gray squirrels, digging up acorns, make similar pits but often leave the peeled half-moons of the husks nearby as well as an oval depression in the bottom of the pit that can be detected by lightly touching with a finger. However, in the absence of additional evidence, there is little to distinguish the digs of skunks from those of other animals that do the same.

Seemingly impervious to stings, skunks often excavate the nests of ground bees. Look for small animal details in the vicinity to distinguish nests destroyed by skunks from bear work. Skunks also dig up turtle eggs, leaving rinds of shells at the hole. However, other mammals like raccoons, foxes and coyotes do the same. The size and distribution of the debris fan at the edge of the hole may help tell whether the work was that of a large long-legged animal such as a coyote or a small, short-legged one like a skunk. When canids dig, they periodically scratch back the debris pile with their hind feet in a violent motion that scatters the dirt over a large area. Also, look carefully at the debris fan from a number of angles; a fresh dig often will show a vague print of the digger.

Dens

Skunks become torpid when the winter snows are soft and deep, moving abroad during thaws or upon formation of surface crust sufficient to support their considerable bulk. Sleeping is usually done in a ground den that the animal may dig itself or borrow from another animal. All too often the selected hole is in the foundation of a house, where the characteristic odor becomes intolerable for the human residents. When the hole is a natural one, fallen leaves may be pulled into the entrance as insulation from the cold outside, a habit shared with other den borrowers like opossums. A sniff around the entrance should tell if a skunk was the resident.

Habitat

Skunks can be regarded as omnivores and so their habitats are varied. Forest, field, shoreline, suburbs and cities all can host a population of these creatures. Unlike its relatives in the weasel family, striped skunks don't climb, and so overhead cover is only valuable to them as a source of food that drops from it to the ground such as nuts, insects, eggs and baby birds. Their only serious predator is the great horned owl, a bird apparently without a sense of smell.

Trackard 20 – Gray Squirrel

Of the half-dozen species of "gray squirrel" in the United States, the two most widely distributed are the Eastern gray squirrel and the larger fox squirrel. All squirrels have similar feet and so distinguishing among them by sign is a matter of print/pattern size and range. All are diurnal and accustomed to humans so that direct observation is easy. The Eastern gray squirrel is presented here as typical of the genus.

As common in city parks as in the wilderness, this small rodent needs little introduction. It is often seen in the fall gathering and burying acorns. In fact, so closely is this animal associated with the seed of the oak that the word "acorn" derives from the Norse word "ekorn," meaning "squirrel." Gray squirrels have a symbiotic relationship with oaks, particularly red oaks, contributing in an important way to those trees' increasing dominance of the hardwood forests in the Northeast. Unlike white oaks, red oaks devote part of their energy to producing bitter tannin in their acorns. When the acorns fall, the squirrels eat the ones from white oaks on the spot and bury the bitter red oak acorns as larder against the coming winter. Later on, preferably when the water-soluble tannin has been leached out by rain or melt, the squirrel digs them up and eats them. The ones that the animal fails to find germinate and become the next generation of red oaks, which grow aggressively and shade out other trees including the few white oaks whose acorns escaped the squirrels' autumn foraging.

Tracks

Much of the Eastern forest has re-matured in recent decades, providing large oaks that, along with other nut-bearing trees, yield abundant mast. As a result, gray squirrels have become abundant as well, and so their tracks are frequently found. In general, gray squirrel tracks resemble those of other members of the family Sciuridae (squirrels, chipmunks and woodchucks), differing mainly in size from the others. For instance, all members of this family show five toes on their hind feet but only four on the front. Also in common with other members of the family, gray squirrels have long toes that they can extend or draw in. On mud, they commonly extend them for traction and support, registering pronounced, long phalangials. This effect is demonstrated in Figure 20.1 C, idealized from a cast in sandy mud in Concord, Massachusetts. In other cases, however, such as on firm, cold surfaces, they may contract their toes and these bones will not show at all. This is the case in Figure 20.1 A and B, recorded in a dusting of snow over ice in Saugus, Massachusetts.

Unlike the hind prints, the front have well developed tertiary pads, a pair of dots that register in most tracks. Although only four toes register in front prints, there is a fifth vestigial toe that may show in very clear tracks as a small blip on the medial side of the tertiary pads. All the front prints presented on Trackard 20 show this vestigial toe. The secondary pads are three dots in a triangular arrangement, like piled-up cannonballs.

The arrangement of the five toes in hind prints is significant to identification. The central three toes are longer than the other two and are commonly grouped together while the medial and lateral toes may be more or less splayed. Furthermore, the central three toes are arranged evenly rather than in an arc. Finally, the lateral two of the central three are often grouped together with a slight space between them and the third. This arrangement may be disguised by irregularities in the surface to which the squirrel adapts with its mobile toes, but it is seen often enough to serve as a distinguishing feature for members of this family and is often helpful in telling right prints from left in obscure patterns. This grouping is most obvious in the left hind print in the upper left corner of Trackard 20 (Figure 20.1).

The hind secondary pads show in tracks as four round or wedge-shaped dots symmetrically arranged in an arc and shaped somewhat like the letter M. Although a couple of tertiary pads are scattered on the animal's heel, they rarely show in prints.

In the Trails and Patterns section at the bottom of Trackard 20, the two rightmost items on the middle line present very typical profiles for gray squirrel hind prints in snow surfaces that do not show individual toes. In such conditions the outline of the print often shows an even leading edge, like an ice-cream cone with a flattened top. On breakable crust the lateral toe is often splayed widely for support and

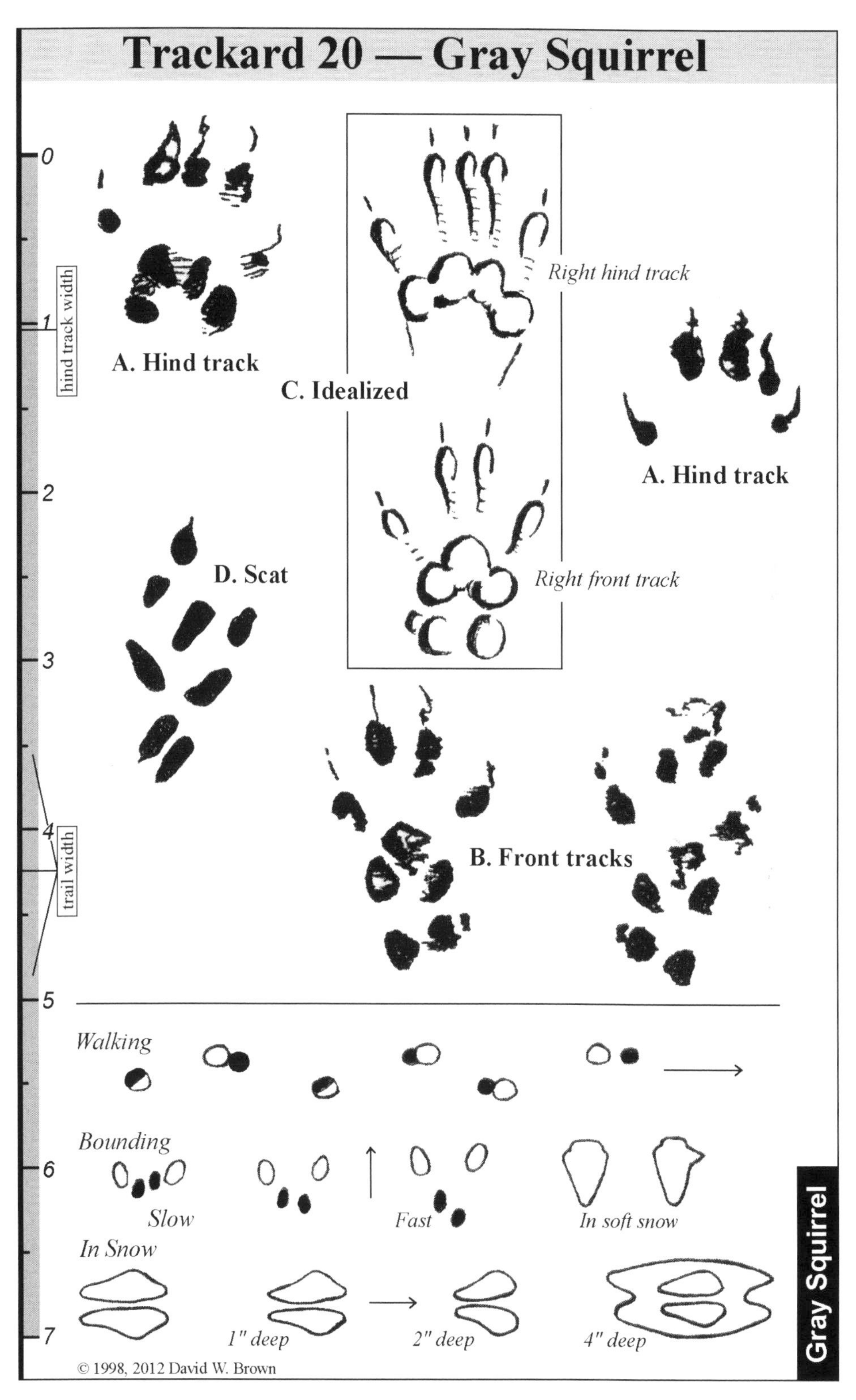

Figure 20.1. Trackard 20 – Gray Squirrel.

will register as a spike on the lateral side of the track as in the rightmost of the two examples. More rarely both lateral and medial toes will do this.

Trails

Although not their usual gait, gray squirrels can diagonal walk like any other quadrapedal mammal, especially when they are foraging slowly on a firm, even surface or when they are trying to avoid breaking through a delicate crust. The resulting pattern is shown in the bottom section of Trackard 20, from a trail on stiffened snow in Saugus. The pattern is a zigzag line of sloppy indirect registrations with a step length of 4–5 inches. Searching along the line of these tracks eventually should show a double registration in which the front and hind prints can be defined and recognized.

The most common gait for squirrels is a bound that leaves a 4X pattern with two more or less even hind prints ahead of and wider apart than the trailing front prints. This inverted trapezoid, or W-shaped pattern, is illustrated by Figure 20.1 A and B combined, which can be seen as a whole by ignoring the inset illustration and the scat. This is the pattern left behind by the squirrel you have seen running between trees in a public park.

These bounds are rolling gaits (see Glossary) in which the hind feet pass by on either side of the planted front and result in the counterintuitive arrangement of hind prints ahead of front. The pattern width for gray squirrels can be as narrow as 3¼ inches in exceptional cases, but is normally between 4 and 5 inches wide, as illustrated actual size on Trackard 20.

In deep, soft snow squirrels and other mammals often use a hop-bound in which the landing front feet pre-pack the snow for the hind feet which follow in right on top or just ahead, depending on the softness of the surface. These patterns often result in a string of triangular patterns in the snow distinctive to squirrels and mice. The line at the extreme bottom of Trackard 20, mostly recorded in Gorham, New Hampshire, shows a series of such impressions in progressively deeper snow, ending with a pattern that shows the animal's body outline. This hop-bound tries for efficiency in difficult conditions, one of the primary characteristics of the winter behavior of any wild animal, by breaking and packing the snow once with the front feet rather than twice with front and hind. The appearance of the gait itself in deep soft conditions is a series of short hops with a high arc and a brief pause between each one while the squirrel extracts its front feet from under the hind and gathers itself for another effort.

Track and Trail Comparisons

Red squirrel: As has been noted, gray squirrels have mobile toes that can be spread or closed and extended or drawn in. Consequently, track measurements can be variable. As a general rule, however, gray squirrel hind prints are at least 1 inch wide across the toes of an adult, even if the toes are closed. The smaller red squirrel, whose prints are identical in shape, usually measures about ¾ inch across these toes. However, accurate measurement of tiny features in the field is often difficult, and there can be a small amount of overlap between the widths of the tracks of reds and grays. Red squirrels also use the same gaits as grays and the resulting patterns can measure with some overlap, as well. However, the pattern width of the familiar 4X bound is almost always narrower with reds, somewhere between 3 and 4 inches across compared to the 4–5 inches usual for grays. In ambiguous patterns, consider habitat and feeding sign, as discussed below, and consider the effects of surface consistency and track age on print size.

Cottontail rabbit: Cottontail rabbits have the same size feet as gray squirrels but usually show an oval print, often with a pointed leading edge, rather than the flat leading edge of squirrels. Having fur-covered feet at all seasons, rabbits also rarely show distinct pads or phalangials or, for that matter, long nails forward of the toe pads.

Rabbits also use the same bounding gait, leaving patterns similar in size and arrangement to those of squirrels. However, squirrels have very mobile feet; a gray squirrel, descending a tree headfirst, can rotate its hind feet around backward for traction with its nails. This rotational mobility often shows in bounding patterns as a tendency for the hind feet to angle outward. Such splaying is illustrated on the middle line of bounding patterns at the bottom of Trackard 20. Rabbits, on the other hand, have relatively rigid feet made for straight ahead speed. Their hind prints tend to be relentlessly parallel. Furthermore, the front prints of the gray squirrel usually register side-by-side or on a slight diagonal, with at least a small space between the two. Front prints of rabbits either register one almost behind the other in a T-shaped pattern to the squirrel's W, or side-by-side and nearly touching. Of course if the trail ends at a tree, any confusion should disappear along with the tracks since few species of cottontails can climb.

Finally, the body outline of a rabbit bounding in deep, soft snow often shows a round tail mark at its trailing edge, while a squirrel's tail, held high overhead, never registers. The resulting squirrel impression shown in the lower right of Trackard 20 is distinctive.

Mink and **weasel:** A mink or a large weasel can leave prints similar in size and with the same number of registering toes as a squirrel. When the medial fifth toe of a mink or weasel registers lightly or not at all, their tracks can look like the four-toed pattern of a squirrel's front foot. If, on the other hand, all five of the weasel's or mink's toes register, the impression can look like the hind print of a squirrel. Furthermore, in mud, a squirrel may dig its long nails straight down for traction, shortening their appearance and increasing the resemblance of its tracks to those of weasel or mink. Where a whole pattern is unavailable and one's identification depends on a single print, as is often the case in summer mud, note that the tertiary pads of a squirrel's front print almost always register. A mink or weasel will show at most only a vague impression in the tertiary area with perhaps a very faint dot on the medial side.

On both front and hind tracks, a mink's toes are arranged in an arc while the central three toes of a squirrel's hind print are even. Furthermore, the common grouping of a squirrel's central toes in its hind print is almost never found in mink or weasel tracks.

When bounding minks and weasels get lazy, as they often do on firm surfaces, the smaller hind prints may register behind the larger front. The resulting four-print pattern can look very much like that of a gray squirrel. However, due to the proportionately wider pelvis of the squirrel, its patterns will show a greater difference between front and hind straddle, resulting in more of a W-shape than the near rectangle of the mustelids. A trail that ends at a tree tells little since all these species can climb, but a trail that ends in a hole in the ice is certainly mink. Gray squirrels can swim, but are unlikely to do so deliberately in winter.

In "deep" snow, minks and weasels also direct register bound, leaving behind 2X patterns similar to those of squirrels, but mostly on a repeating slant rather than even, as with sciurids, and with relatively longer inter-patterns distances.

Scat

Gray squirrel scat is usually more or less ovoid, often pointed at one end and blunt at the other. The collection in Figure 20.1 D is from Petersham, Massachusetts, in winter. Accurately measuring such small items in the field is difficult, but the maximum diameter should be about 4–5 millimeters. Although there is a lot of overlap with scat of red squirrels, differences in preferred habitat as well as feeding patterns discussed below may help in the accurate identification of the scat.

Sign

A totally exposed acorn meat with only the top, the part with the least tannin, gnawed off and peeled shell crescents nearby is definite sign of squirrel work. Gray squirrels tend to eat where they find food rather than to move food to a feeding perch. Furthermore, rather than creating a concentrated cache, they store individual nuts in separate places underground. They are messy eaters, tending to leave debris scattered around the feeding site, often at the shallow hole in the ground from which they have extracted the nut they stored there earlier. Gently exploring the excavation with a finger may reveal the smooth oval of the removed nut. They have such exquisite sense of smell that even in deep snow, gray squirrels can locate individual buried nuts and dig down to them, leaving characteristic holes as deep as 2 feet.

An interesting marking behavior of gray squirrels can sometimes be seen on the bark of tree trunks and branches where the animals progressively chew up the surface of the bark. This gnawing lengthens into a long roughened area which is probably scented from one or another of the animal's glands. This scenting may tell other squirrels in the area information about the creator, such as fitness and breeding condition.

Sign Comparisons

Red squirrel: In determining the identity of the feeder from the debris it leaves behind, remember that red squirrels generally feed at a favorite elevated perch near their cache and leave at least a small "midden" pile behind. Grays tend to feed where they find food. Any individual, thoroughly destroyed nut or acorn shells in a hardwood or mixed forest are a pretty good indication of gray squirrel.

Chipmunk: Chipmunks also eat nuts such as acorns. The resulting debris is often in the form of a small collection of shells on top of a stone wall or other elevation near the animal's burrow system. Acorns will be less thoroughly disassembled than is the case with squirrels, often showing gnawing down from the blunt end to about half-way (see Trackard 23 – Chipmunk, Figure 23.1). No animal does more work to get its food than it must. The chipmunk, with its smaller head and mouthparts, doesn't need to peel an acorn totally to extract the meat. In distinguishing the work of squirrels, chipmunks and mice, look among the debris for the nut with the smallest opening through which the contents have been successfully removed and then imagine the size of the mouth parts and head that performed the extraction.

Den

Gray squirrels usually den in bad weather in hollows in trees that may be used as birthing chambers as well. They also construct outside "leaf nests," which are scraggly accumulations of twigs and leaves. These are a common sight in residential areas where available holes in shade trees are insufficient to house a proportionately large population of squirrels. Younger forests without many natural tree holes also may provoke such construction. In woodlands these leaf nests may be used by other animals, including fishers, as resting and caching locations. They may also be used by large birds as the foundation of their own nest. Although red squirrels also occasionally build outside nests, theirs are smaller, neater and more spherical than the unkempt nests of grays.

Habitat

Gray squirrels inhabit deciduous and mixed deciduous/coniferous woodlands where there is a crop of nuts, especially acorns, available for winter food. Note that in dense conifers, however, red squirrels are likely to be the only squirrel present. Suburbs and cities with streets lined with hardwoods as well as public parks also provide the nut crops grays require. In addition to nuts, squirrels also eat other food in season such as fruit, fungi, birds' eggs and even the cambium of downed trees.

Trackard 21 – Weasel

Of the members of the family Mustelidae the smallest are called weasels. Three species of these little predators inhabit North America with two, the long-tailed and the short-tailed, ranging into New England. In winter, in the northern areas, these weasels molt their fur to white with, on the tail of the larger two species, a black tip. In this season they are sometimes called "ermine," although the term is also used at all seasons for the smaller, short-tailed weasel in Britain and increasingly in America. These ferocious little mammals are mainly nocturnal, where white camouflage provides little hunting advantage, and, with their long, slender bodies and short legs, are adapted to hunting for subnivean rodents in the darkness of tunnels. One might wonder, then, why the species expend precious energy changing to white fur every fall and back again to ruddy fur in the spring. The answer is that white fur is hollow and provides better insulation than dark fur, the fibers of which are filled with melanin granules that make them more conductive. Since a long, thin body is not the best shape for heat conservation, weasels in winter need all the help they can get. In addition, weasels are sometimes mistaken for prey species in the day by hawks and at night by owls, to the mutual sorrow of bird and mammal since lore has it that a weasel caught in the talons of a bird of prey may kill its attacker while dying itself. So their seasonal pelage may be for protection from other predators as much as for heat conservation.

Tracks

The only discernible difference among the prints of these smaller members of the weasel family, at least in winter, is size (Figure 21.1).

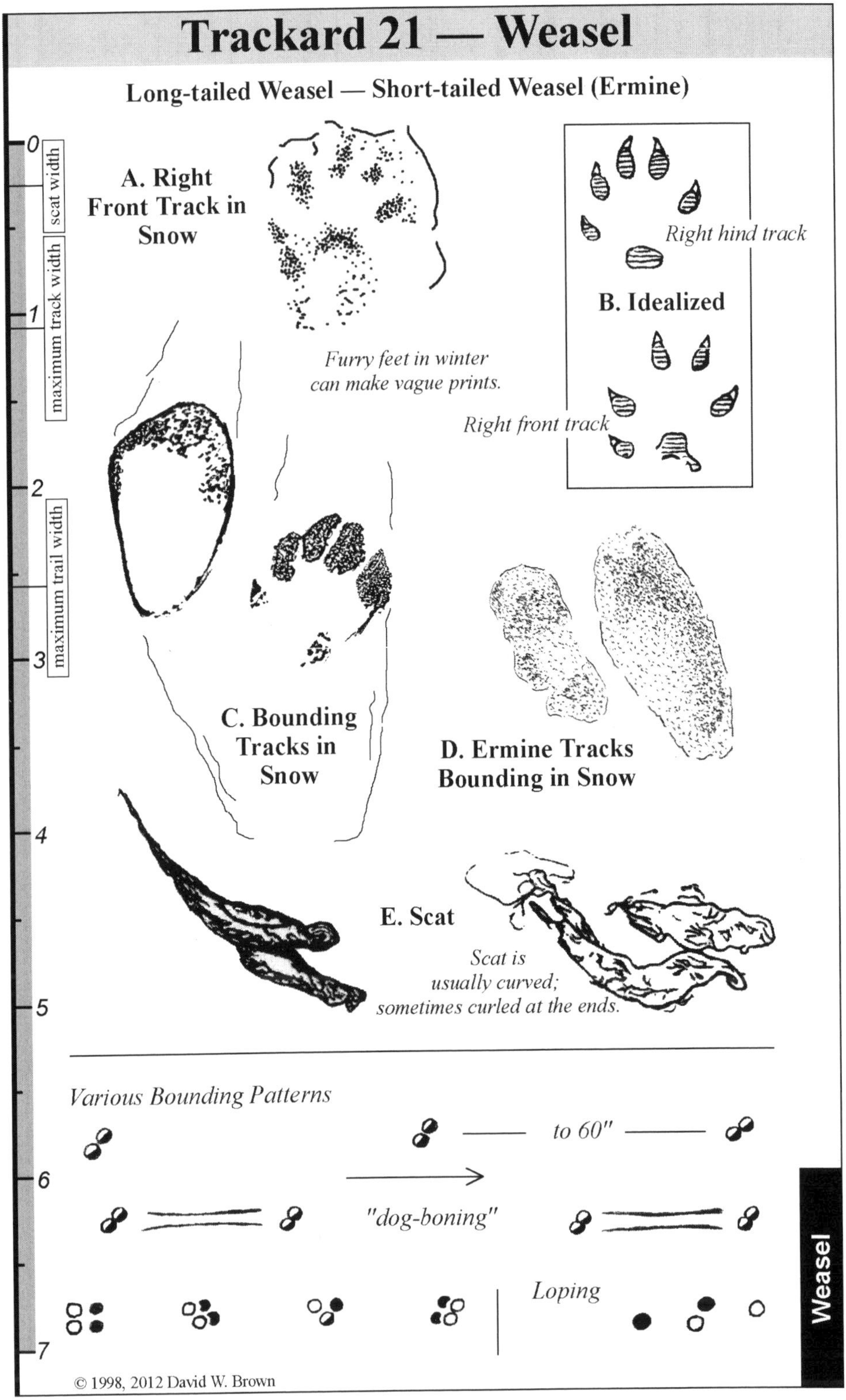

Trackard 21 — Weasel

Long-tailed Weasel — Short-tailed Weasel (Ermine)

Figure 21.1. Trackard 21 – Weasel.

Unfortunately for the tracker, all weasels are sexually dimorphic, with the males about a third larger than the females. Furthermore, weasel species seem to gain and decline in size corresponding to the population peaks and crashes of various sized prey animals. As a result there can be a lot of confusion as males of the smaller short-tailed species, for instance, overlap with long-tailed females and the tracks of the long-tailed males approach the size of female minks. Generally all these animals show mustelid type prints so their features alone will not help to tell the difference. Figure 21.2 H shows the feet of a road-killed male long-tailed weasel. Note the usual mustelid characteristics: five toes arranged asymmetrically, with the medial fifth toe retarded. The secondary pads show lobes corresponding to the toe pads including a small fifth lobe on the medial side. On the front print a well-developed tertiary pad on the lateral side appears. All of these characteristics are common to both weasels and minks. Both the front and hind feet in this illustration measure about ¾ inch across, but it must be remembered that a foot is not a print. Weight bearing and muscular control can be expected to spread the tracks, especially the front, and the usual weasel bounding gait with its high impact will fatten and round the pads. The result is that one should add about ¼ inch to the relaxed foot width of the dead weasel. Thus 1 inch probably represents the upper end of the track size for this and other species of small weasel.

However, the problems don't end there. When vole populations crash and lagomorph populations are high, weasel species may increase their size to accommodate the larger prey. This increase can occur quite rapidly given the short life-span and correspondingly high reproductive turn-over of weasels. When hares crash and voles are abundant, on the other hand, the species can quickly reverse the process and select for smaller size accordingly. All of the illustrated tracks in this chapter are of long-tailed weasels except the amorphous bound pattern in Figure 21.1 D, which was made by an ermine. The differences among the short-tailed illustrations show some of the variability of track size for a single species. It should also be noted that the road-killed weasel's feet were photographed in summer when they had less interdigital fur than in the winter when extra fur as well as registration in soft snow invariably enlarge track size.

In differentiating front from hind prints, note that, because the front feet support the weight of the head, the front toes tend to splay

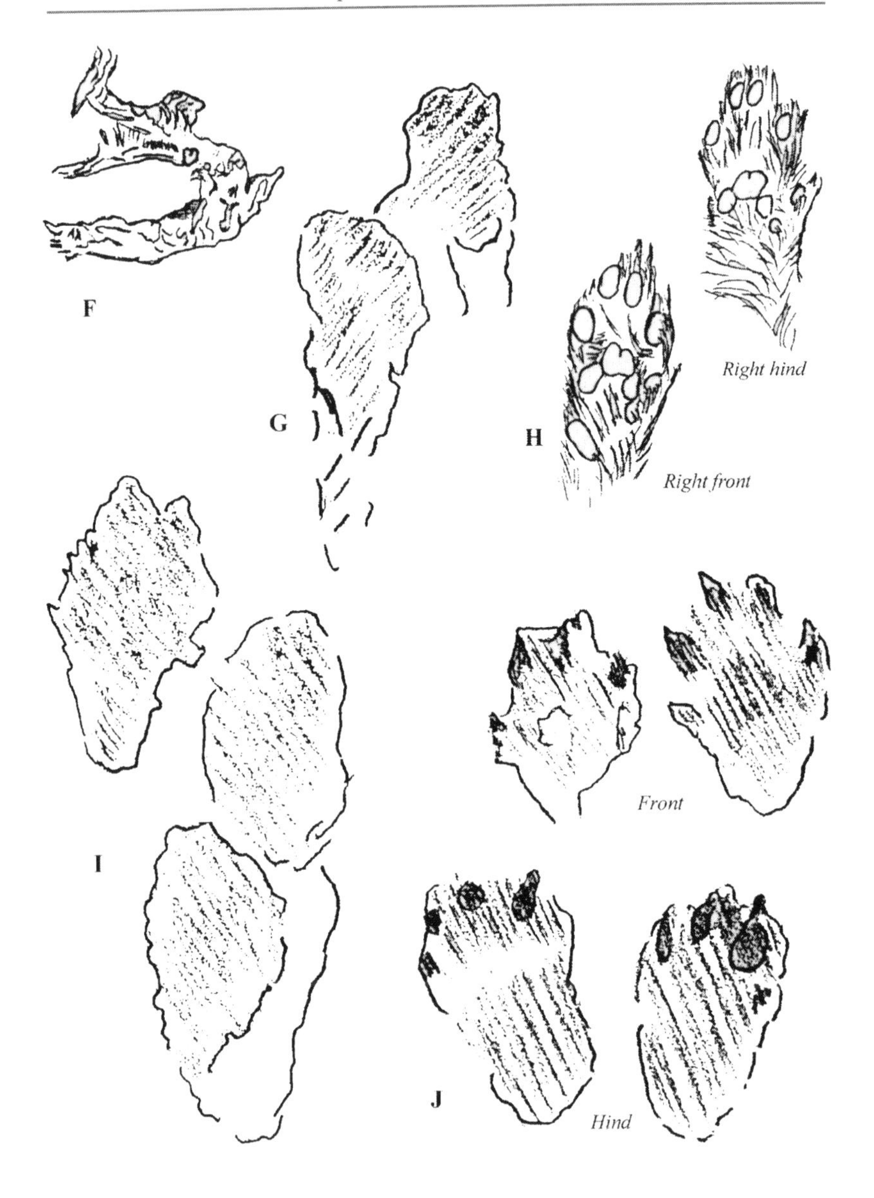

Figure 21.2. Weasel patterns, scat and feet, about 2/3 actual size.

F. Feather scat. Saugus, Massachusetts.

G. 2X bounding pattern in soft, dry snow. New Salem, Massachusetts.

H. Feet of male long-tailed weasel in summer. Carlisle, Massachusetts.

I. A three-track (3X) bounding pattern in dry snow.

J. Slow bound pattern showing heel of hind prints. Princeton, Massachusetts.

more than the hind. Bisecting the impressions of the two central toe pads along their long axis shows more or less parallel lines for the hind and splayed lines for the front. This effect can be seen most clearly in figures 21.1 B and 21.2 J. The former is idealized from several photographs and the latter was found in a gap of a stone wall along which the animal had been traveling.

Trails

Weasels must run a very high metabolism in winter to keep up with the high rate of heat loss resulting from their body shape and small size. At rest, they conserve heat by curling up in a ball and by using the insulation of an appropriated burrow or hollow. Abroad they must keep moving and so are extremely active hunters. This is reflected in the nervous, darting nature of their trails, as they follow their senses in the direction of any potential prey. A high metabolism must be fed often, and the short life expectancy of many weasels is the consequence of lack of hunting success. Trails in soft snow often disappear as the animal plunges downward in search of voles at ground level only to reappear a few yards farther on. Not only may the weasel meet with hunting success during such plunges, but it also gets to use the insulation of the snow to reduce heat loss.

Weasels' most common gait is a bound, the high arc of stride of which lets these little, short-legged animals vault over debris on the forest floor, see over dense vegetation and sense potential prey in their vicinity. To an observer, the appearance of this gait is peculiar; the front of the animal seems to come down and bounce up before the hind comes down and then does the same. The visual result is a floppy motion, like a rag being shaken on the end of a stick. Even when the forest is smoothed out by snowfall, weasels continue to use the bounding gait where the surface is at all soft because its extra verticality allows them to clear the intervening snow.

Weasel prints are so light that they are usually found only in the winter. In soft and "deep" snow, which for a weasel means a track depth of as little as an inch, the bounding pattern becomes a series of 2X direct registrations on a repeating slant with a narrow trail width and a long inter-pattern distance. See the uppermost pattern in the trails section at the bottom of Trackard 21 (Figure 21.1). The energy generated by this gait is amazing. The animal is like a spring, vaulting easily 5 feet at a time. (In one remarkable instance I found where a long-tailed weasel, to avoid exposure to a chilling wind in an opening in a pine plantation, had made three successive bounds of 7–9 feet!) 2X track patterns from bounding trails are shown in Figure 21.1 C and D and in Figure 21.2 G. In Figure 21.1 C, from the Quabbin region of central Massachusetts, the depth was about an inch, and the snow consistency was damp enough to register pads. Figure 21.1 D, on the other hand, from Barre Falls in Massachusetts, shows a pattern from the trail of an ermine in dry soft snow. Even though no pads were evident, a trail of patterns in repeating slants identified it as a weasel, and both the narrow trail width and small track size refined that identification to an ermine. A similar direct registering pattern in Figure 21.2 G shows a more obvious slant. Here too the snow was so dry and fresh that only the outline of the tracks showed.

When the snow surface is firm and the efficiency of direct registering is not needed, bounding weasels may get a little lazy and begin to double register, with one or both hind feet landing off-center from the front. Figure 21.2 I shows a 3X pattern of this latter sort. A variety of this kind of pattern is shown on the bottom line of the trails section on Trackard 21, as well, including several where the animal overstepped or understepped. Occasionally one may find a true 4X bounding pattern of single registrations such as in Figure 21.2 J. Here the tracks were recorded after a thaw, in thin snow over frozen melt-ice over bare ground. The weasel had been following along the top of a stone wall and had jumped down into a gap where it gathered itself to leap back up onto the next section. In this case the outline of the heel area of the hind feet can be seen.

In firm snow conditions and especially in open areas, weasels may opt for a faster loping gait. The lower arc of stride means greater ground-speed across potential danger areas where they might be intercepted by a patrolling hawk, and the smooth, firm surface does not require the waste of energy implicit in the greater verticality of a bounding gait. A typical loping pattern common to all mustelids is presented at the extreme bottom right of Trackard 21. In this case crusted snow on an abandoned road in New Salem, Massachusetts, allowed the weasel to lope along for nearly a quarter of a mile using a variety of such "1-2-1" patterns.

In addition to the short-tailed weasel (ermine) leaving tracks and trails of smaller proportions than its larger cousin, it also often shows a distinctive pattern in its 2X bounding gait, which is illustrated on Trackard 21 in the middle line of the trails section, recorded in Saugus, Massachusetts. Short-tails often alternate long jumps with short. In soft snow the short jumps may be connected by a trough, showing that the animal reduced its verticality with every other stride. The combination of the slant-2X tracks connected by a trough is often referred to as a "dog-bone" or "dumbbell" effect.

Not only do otters and minks slide in snow, but weasels do as well. While otters clearly do this for play, however, the smaller mustelids with their high metabolism and more precarious existence seem to use sliding as an energy-saving measure, doing it only in snow conditions where it is an efficient way of getting down a slope. The resulting trough will be about 3 inches wide, compared to the usual 4–5 inches for minks and 7–9 inches for otters.

Track and Trail Comparisons

Red squirrel: The difficulties of distinguishing the tracks and trails of closely related mustelids have been discussed above. Another common difficulty occurs with red squirrels, which have track sizes in the weasel range and leave some similar patterns. Where both weasel and red squirrel direct-registered in a bounding gait, look for the repeating slant of the weasel to the even arrangement of tracks in the squirrel's trail. When a squirrel is 2X bounding, the inter-pattern distances will also be short compared to a weasel's in the same conditions.

When a weasel understeps in a bound, as they often do on firm surfaces, the splay-toed front feet and the smaller hind can be arranged in a 4X pattern that looks remarkably like the familiar bound pattern of a squirrel with its larger hind feet advanced. If the prints are so vague that toes can't be counted and hind feet told from front, note that weasel trails tend to have a narrower straddle relative to a longer inter-pattern distance than the trails of squirrels. Also, a trail that winds and darts over the surface, investigating hiding places is almost certainly the trail of a weasel. In winter, red squirrel trails tend to go directly from den to cache or other food source and back. Red squirrels have neither dense underfur nor the weasel's high metabolism, either or both of which are needed to spend a lot of time out on the surface of the snow.

Scat

Weasel scat is normally easy to identify. It is very small and generally composed solely of the fur or feathers of prey animals; a tail of this material will appear at one or both ends. The scat, which frequently has a braided or twisted appearance, is often formed in an arc that results from deposition by a short-legged animal whose anal opening is very close to the ground. The scat in figures 21.1 E and 21.2 F, all from Saugus, show variations. The scat on the left of Trackard 21 was composed of fur such as one might find on a shrew or mole and was fresh enough that it still contained mucous that darkened its appearance. The scat on the right was deposited in the middle of an unused asphalt road. At first glance it looked like a scrap of discarded pocket lint, but a closer inspection showed the tails of hair and the braiding effect. This scat showed no dark mucous; the weasel may have revisited an old cache that previously had been reduced mostly to fur. The scat shown in Figure 21.2 F, which was gray and composed of feathers, was deposited on moss along the bank of the Saugus River.

Sign

When weasels are in a feeding frenzy or are otherwise excited, they may emit musk, the skunk odor characteristic of all mustelids. This odor may linger for some time around a kill site. Since weasels are good climbers, this odor at a ravaged bird's nest in a bush or tree will indicate the weasel's responsibility.

The long, narrow weasel shape, however successfully it allows these predators to hunt down holes and tunnels, also dictates a very small stomach. Sometimes they kill more than they can consume at the moment in a spasm of surplus predation. Several uneaten or partly eaten carcasses at a communal rodent nest in winter, for instance, especially with a musk odor associated, will indicate weasel predation.

Habitat

In some guides, habitat is used to distinguish short-tailed weasel trails from those of long-tailed and short-tailed weasels from minks. It is sometimes suggested that long-taileds favor bottomlands more than the uplands preferred by short-taileds and that mink trails follow the edges of water and are not normally found in uplands. Experience shows, however, that all three species may be found at one time or another in bottomlands or uplands, whatever their supposed preferences may be. One cannot tell in any given instance whether the trail one finds is the rule or the exception. Following the trail for any distance should distinguish mink from the others. Being well adapted to water, minks found in the woods will usually be on their way from one wetland to another; following the trail for a ways will show the connection. This is not always true, however. Occasionally, minks will hunt well away from water, and in late winter males roam widely in search of mates. Few rules are applicable 100% of the time in tracking.

Both species of weasel will show hunting rather than traveling behavior in their upland trails, zigzagging here and there, wherever their nose or curiosity leads them in search of a potential meal. Their trails often follow stone walls as the animals search for small rodents that live in the crevices. From this elevated position they can also better see or otherwise sense potential prey nearby on both sides of the wall or log. Or they may run along a brushy woodline next to a field where cottontails might be hiding in their forms or voles in their tunnels.

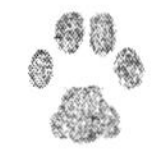

Trackard 22 – Red Squirrel

Red squirrels range across the northern United States and Canada, replaced on the West Coast by the similar Douglas's squirrel. Western versions show a browner coat than the eastern variety. Smaller than the gray squirrels whose range they overlap, red squirrels make up for it by their aggressiveness. They are highly territorial, normally permitting the close approach of others of their species only at mating time or at a concentrated food source such as a bird feeder. They are quite excitable and feisty on their territory, buzzing and chattering hoarsely at you once they discover your presence. Unlike gray squirrels, reds store their winter food in a concentrated larder, gathering green nuts and cones into an underground chamber in the late summer and fall. This chamber is often dug into the rotting root structure under a stump. Also unlike grays, when red squirrels find food, they tend to move to a feeding perch to eat rather than consuming it on the spot. Red squirrels can swim and will do so for long distances, across a pond for instance, if a food source warrants the effort.

Tracks

Like other squirrels and most rodents, red squirrels have four toes on the front foot and five on the hind as shown in Figure 22.1 A, cast in mud at Stoneham, Maine. The toes may be spread or closed, extended or drawn in. When they are extended, phalangials often show in the prints. But when they are drawn in, only the toe pads register, appearing as ovals with the long axis in the direction of travel. Nails are often extended and prominent in very clear prints. In these respects and all others except size, red squirrel prints resemble those of the larger gray squirrel and smaller chipmunk.

The hind print of a red squirrel shows the typical squirrel arrangement of three long central toes. The medial and lateral toes are somewhat shorter and tend to splay more, causing the track width to vary a little with the amount of this splay. The width is generally under an inch and usually around ¾ inch across the five toes compared to the usual 1 inch or more for a gray squirrel. As with most other sciurids, the three central toe pads are even and often grouped, with the lateral two slightly closer together. The four secondary pads in the hind print, which register as an arc of four balls or wedges behind the toe pads, are arranged more or less symmetrically. No tertiary pads register in the print although the outline of the heel area may show, especially in snow.

The front print of a red squirrel is smaller than the hind, has four toes and shows only three protuberances in the secondary pad, arranged pyramidally in the horizontal plane like a pile of cannonballs. Tertiary pads normally register and appear as two balls to the rear of the secondary pads. In very clear prints the vestige of a fifth toe can be discriminated as a tiny blip on the medial side of the tertiaries. This shows most clearly in the bottom front print in Figure 22.1 A.

Trails

The normal gait for a red squirrel, like others of its family, is a bound in which the hind feet register ahead and wider apart than the front feet. This four-print (4X) pattern is shown in Figure 22.1 A, actual size. Several variations appear in Figure 22.2 D-G , all photographed in shallow snow at Passaconaway. As the speed of the bounding animal increases, pattern length stretches out as well.

When snow depth reaches 8 inches or more, red squirrels take advantage of its insulating effect by tunneling under it to connect den with cache. By using these tunnels a red squirrel not only protects itself from the wind chill and frigid temperatures on the surface overhead but also conceals itself from predators such as the fox, coyote and hawk. These tunnels are most often found in the thaws of late winter and spring when they are exposed as trenches about 4 inches wide with many tracks coming and going.

In soft snow where the animal is sinking in more than a couple of inches, red squirrels commonly hop-bound, a gait in which the front feet pre-pack the snow for the hind, which follow in on top of or slightly ahead of them. The resulting pattern is shown at the extreme bottom right of Trackard 22. This pattern shape is a signature for squirrels, chipmunks and mice, differing among them only in size.

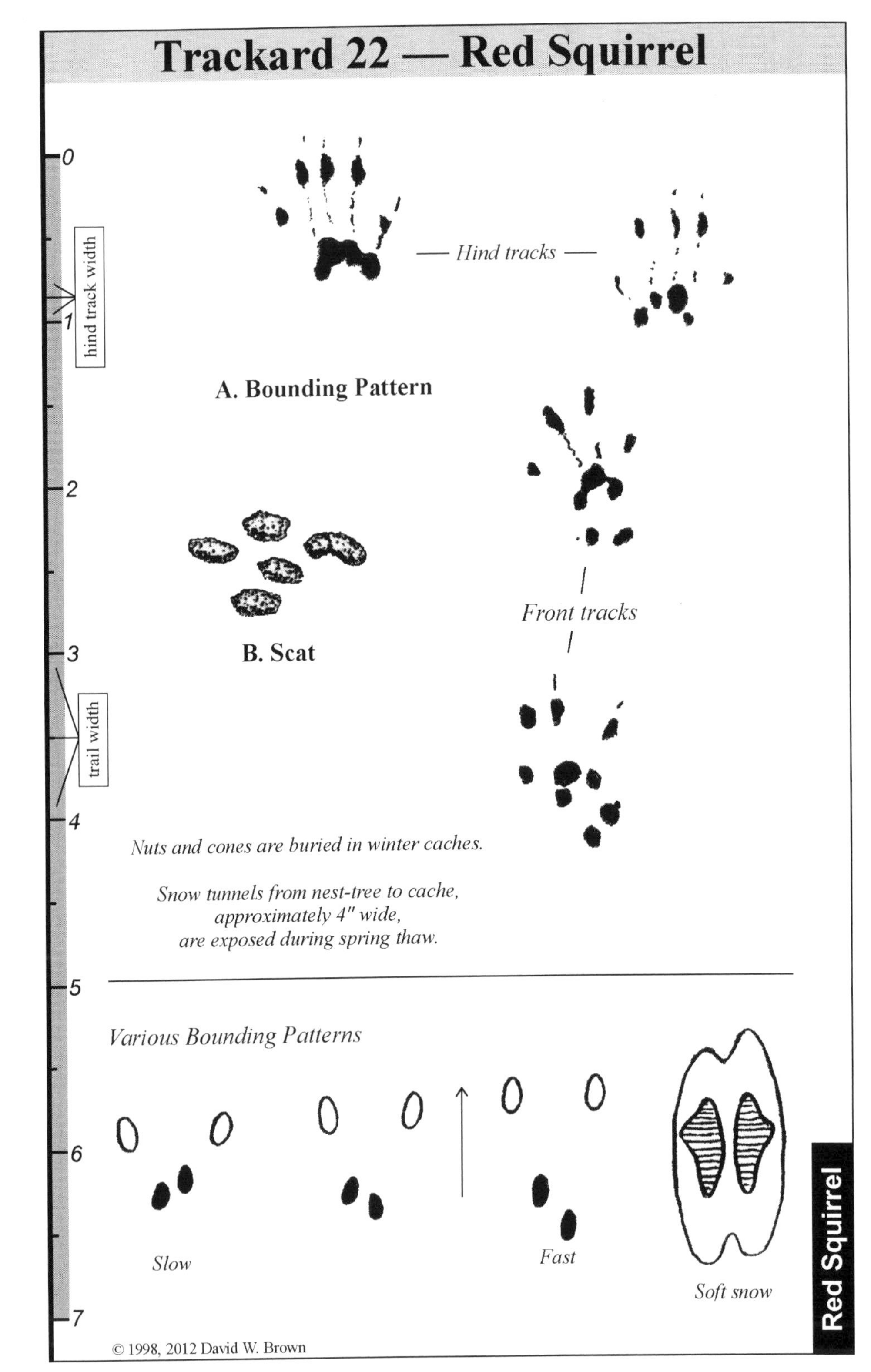

Figure 22.1 Trackard 22 – Red Squirrel.

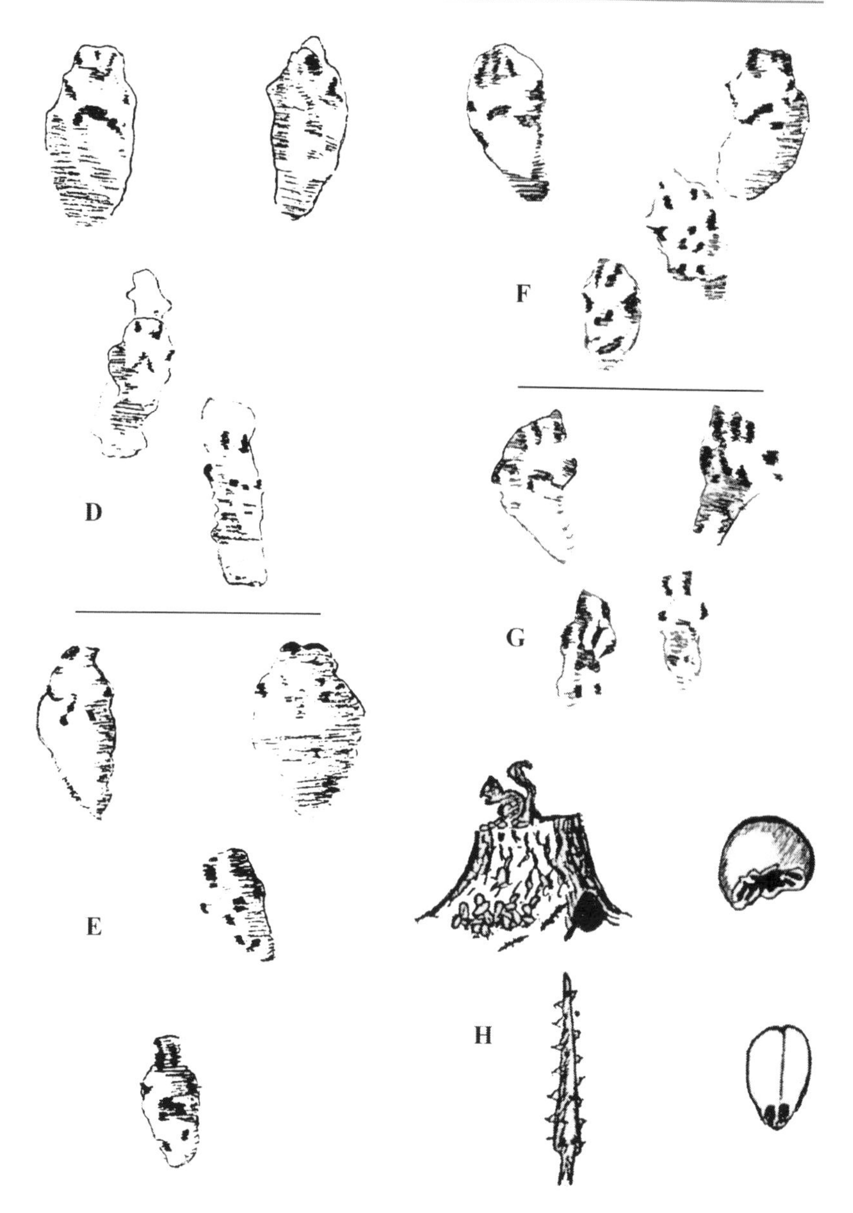

Figure 22.2. Red squirrel patterns and sign.

D, E. 4X bounding patterns at speed in shallow snow, about 2/5 size.

F, G. Same gait at slower speed. Note compression in direction of travel.

H. Clockwise: stump feeding perch with midden and cache entrance below; opened hickory nut; pine cone scale with seeds at base; pine cone core or "cob" stripped of scales.

Track and Trail Comparisons

Gray squirrel and **chipmunk:** The 4X trapezoidal bounding pattern is the usual shallow surface pattern of all squirrels and chipmunks as well as most mice. The main distinction is size. Red squirrel pattern widths (hind straddles) are usually between 3 and 4 inches. All of the patterns presented in Figure 22.2 measured 3¼–3½ inches across the hind prints. (Figure 22.1 A shows a slightly reduced pattern width caused by the animal changing direction.) Gray squirrels generally measure 4–5 inches in hind straddle and chipmunks 2–3 inches, although with very flexible hind legs designed for tree climbing, red squirrel straddles can overlap with both species. In distinguishing among them, a number of clues need to be looked for and a judgment made on the sum of the evidence.

In pure conifer stands, any squirrel pattern of appropriate size is almost certainly a red squirrel. In hardwoods or mixed forest, grays are the more usual squirrel, but here reds may invade as well, especially if there is a concentration of nut-bearing trees. Chipmunks frequently forage in weedy openings near stone walls and treelines where they often move slowly and leave patterns of very short length, with the front prints between rather than behind the front. Red squirrels are more arboreal and tend to avoid woods openings as danger areas. When they do enter an opening, it is usually to retrieve a fallen cone and so the trip will be faster, leaving longer pattern lengths and a more purposefully direct trail than is common for foraging chipmunks.

While splaying of hind feet in bounding patterns is common for all squirrels, a slight toeing-in of one or both front prints in snow is a habit I see more often with red squirrels than grays. Most of the illustrations in Figure 22.2 show this effect. Careful examination shows that this toeing-in is something of an illusion, having more to do with the entry angle of the foot in snow rather than the angle of the print itself. As a result, it will not appear in a very shallow track. However, it is an illusion common enough with this species that I regard it as a signature clue where other distinguishing traits are absent.

In winter chipmunks hibernate below ground; their patterns are likely to be found only during a thaw. Gray squirrels, on the other hand, are abroad throughout the winter, excavating buried acorns. These digs are in scattered locations, quite unlike the red squirrel, which relies on a concentrated cache for its food. Red squirrels, which do not add extra fur in the winter to any great degree, are vulnerable to cold. When they travel on cold days, it is usually between den and cache, and they do so as quickly as possible, leaving straight trails and long pattern lengths with their 4X bounds.

Long-tailed weasel: Comparably sized weasel prints may show five toes, like a red squirrel's hind print, but without the even arrangement or differential grouping of the central toes. In addition, weasel secondary pads are asymmetrical although this may not be apparent except in clear prints. When the larger weasels use a bounding gait, the tracks will be on a repeating slant, rather than perpendicular to the direction of travel, and with a narrower trail width and much longer inter-pattern distance than the similar hop-bound pattern of a squirrel. For other discussion, see Track and Trail Comparisons in the chapters on weasel, mink and gray squirrel.

Scat

There is little to distinguish the scat of red squirrels from that of gray squirrels. Gray squirrel droppings might be expected to be larger, but the size difference in items so small can be very hard to discern. Individual droppings are normally granular, ovoid, sometimes pointed at one end and blunt at the other. Their presence in the vicinity of a midden of cone scales (see below) will be diagnostic.

Sign

Rather than feeding where they find, as gray squirrels do, red squirrels normally gather a nut or cone and move to the nest or, in mild weather, to a feeding perch to peel and eat it. This perch is often a dead stub on the lower trunk of a pine, spruce or fir tree. In this position the animal can better sense the approach of a predator. If a fisher runs up the back of the tree, the sound of its nails may give it away. If a hawk tries to make a pass, the trunk will interfere with its flight, and the squirrel's elevation protects it from the red (but not the gray) fox or coyote below. These feeding perches may be used so repeatedly that a pile of cone scales, called a "midden," may accumulate on the ground below. A sure sign of this species, middens range from as little as a cup or two of scales to huge piles. Sometimes red squirrels dig caching chambers in the soft root area under a rotting stump and use the stump itself for a feeding perch as shown in Figure 22.2 H. If danger is sensed the squirrel can dive into its caching chamber or run up a nearby tree.

Other than by location, distinguishing feeding sign on acorns and nuts from similar feeding by gray squirrels is difficult. Since both have fairly large mouthparts, they both must disassemble an acorn thoroughly to get at the meat. And hickory nuts with large ragged openings may be the work of either species. Chipmunks usually are obliged only to chew off about half of an acorn shell to extract the kernel, and mice make even smaller openings. Flying squirrel work on a hickory nut is very smooth, with the edge of the opening appearing beveled. One oddity that I have observed about red squirrels and hickory nuts is that the animal sometimes halves them neatly at an invisible seam before storing them in a cache. Presumably, this makes them easier to open later for actual feeding.

The midden pile that accumulates under a feeding perch of a red squirrel will be composed mostly of cone scales and cone cores or "cobs." The little animal can strip the scales from an entire white pine cone in about a minute, extracting the two small seeds, or "pine nuts," hidden at the base of each scale. Once these seeds are removed, the scale is dropped and added to the pile below. When the cone is completely stripped, the cob is dropped as well.

A mushroom left on the ground decays rapidly to an inedible mass. To prevent this and prepare a mushroom for winter storage, red squirrels often wedge it in the lower dead branches of a tree where it will dry without deteriorating. The fact that I often find these in dead pine branches has suggested to me that this tree may have bacteria-suppressing qualities that also retard disintegration.

Dens

Red squirrels den for birthing and for protection from hawks, owls and cold. In mature forests the den may be in a tree hole, while outside nests are more common in a young forest without many natural cavities. An outside nest of this species is more spherical and neatly constructed than the scraggly leaf nests of gray squirrels. Any globular nest of appropriate size in a conifer stand in the Northeast can safely be attributed to the red rather than the gray squirrel.

Habitat

While the pelage of a gray squirrel neatly matches the color of the bark of an oak or other hardwood, red squirrel fur is the same color as both dead conifer needles and dead hardwood leaves on the forest floor, affording the little animal some protection in both habitats. It spends little time in openings where it blends poorly with green or dead grass, and it seems to avoid brushy areas such as clearcuts as well, both frequent chipmunk habitats. Since it feeds mainly on cones, nuts and other tree products, it is decidedly arboreal. In the dense spruce-fir forests of the far North it replaces the gray squirrel entirely.

When an artificial food source such as a bird feeder is located in a red squirrel's range, its normal behavior may be modified: it will stay out in the cold longer, will scamper across openings to get to the seed, show foraging behavior under the feeder and even tolerate the presence of others of its species, however irritably. Under such circumstances, distinguishing the tracks and sign of this species from other squirrel visitors can be challenging.

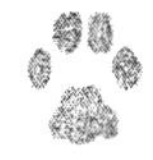

Trackard 23 – Chipmunk

Seventeen species of chipmunks, small squirrel-like animals, inhabit various parts of the United States. Most are about the same size, and the West has a disproportionate share of them. The one with the widest range is the Eastern chipmunk, the information for which is presented here. The next most widely distributed is the somewhat smaller least chipmunk of the western states.

Eastern chipmunks, with a ruddy blond pelt and stripes on their backs, are a common sight to anyone who tramps the New England woods. They live in and above an elaborate tunnel system that they defend as territory from others of their kind. So boisterous and aggressive is this defense that chipmunks chasing each other may be oblivious to human observers who may mistake this determined defense for play.

Tracks

Chipmunks go into extended torpor in their tunnels for most of the winter. When their prints are seen in snow, it is usually during a prolonged thaw or after an early spring flurry. At other seasons, their patterns may be found in any mud near woods such as streambanks or tire ruts on a dirt road.

Hind prints are normally digitigrade, that is, only the toes and the secondary pads register in the track as shown in Figure 23.1 A, cast in mud at Templeton, Massachusetts. Just like squirrels, the three central toes are longer than the medial and lateral ones, and there is a strong tendency for the lateral two of the central three to be grouped. The phalangials may or may not appear in the print depending on the depth of the tracks and whether the animal flexed its toes. The secondary pads on the hind are four round dots arranged in an arc shaped roughly like the letter M. Tertiary pads do not show in hind prints. In these and all other respects the hind and front prints of chipmunks resemble those of squirrels, only smaller.

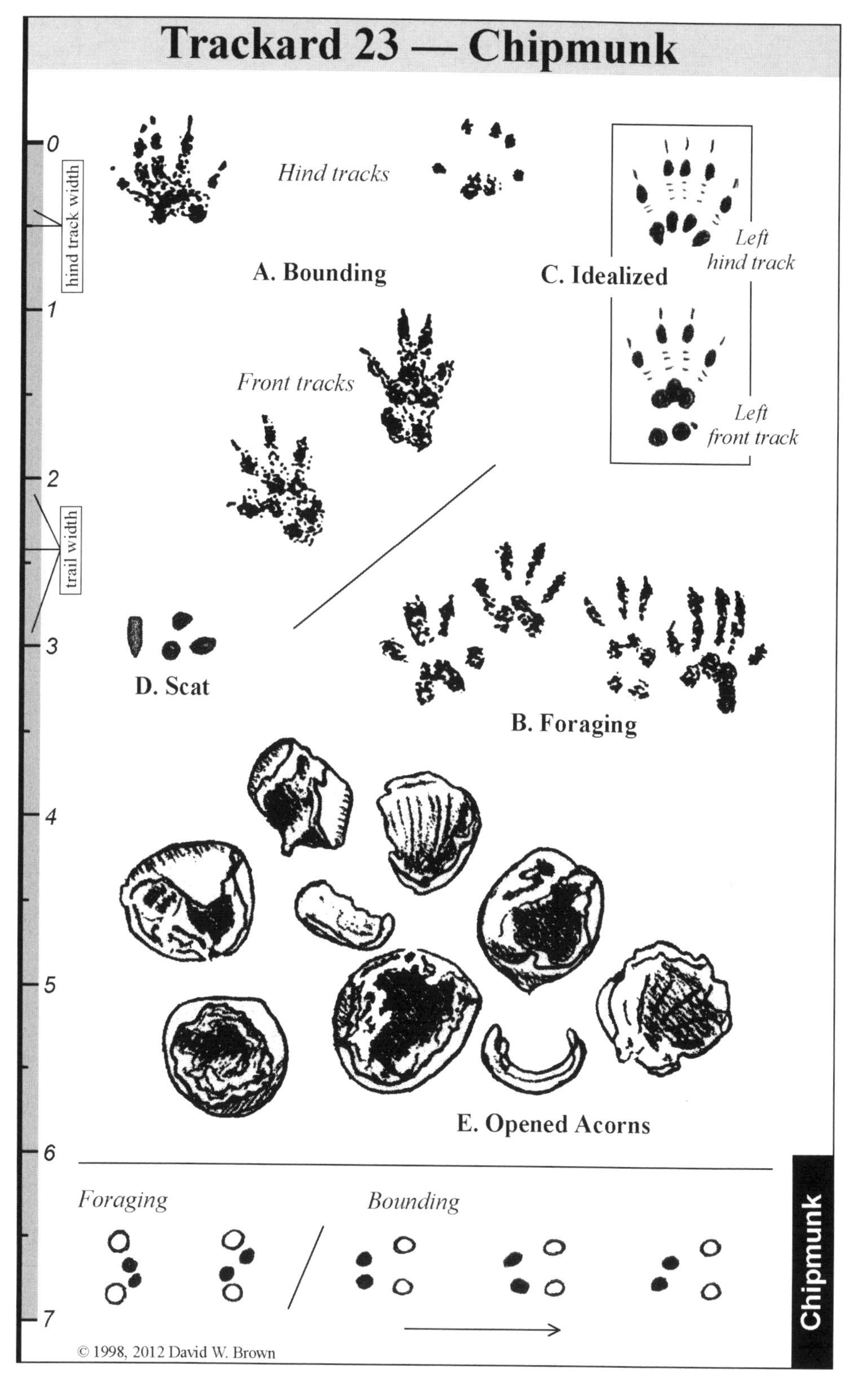

Figure 23.1. Trackard 23 – Chipmunk.

Front prints show only four toes, with the vestigial fifth represented by a bump on the medial margin of the tertiary pads. This is shown in Figure 23.1 C and can barely be seen on the right front print in Figure 23.1 A. There are only three secondary pads, represented by a triangle of dots. Tertiary pads normally register in front prints, appearing as two dots to the rear of the secondary pads as is the case on three of the four front prints presented in Figure 23.1 A and B.

Trails

Like most other small rodents, chipmunks travel by bounding much more than by walking. Figure 23.1 A shows the resulting trapezoidal pattern with hind feet advanced and paired while the front prints are retarded. Also similar to other small members of its taxonomic family, Sciuridae, chipmunks have narrower chests than pelvises so that the hind straddle is always wider than the front. The bounding pattern is mainly seen where a chipmunk is escaping danger or chasing another of its species out of its territory, both very common behaviors for this quick-tempered little animal.

Figure 23.1 B, from a cast in mud at Saugus, Massachusetts, shows a common foraging pattern, a longitudinally compressed version of the bounding pattern. Here front feet are even with or slightly ahead of the hind. This distribution denotes an animal moving slowly, searching in front of its nose for things to eat. When it finds something, it usually feeds on the spot or moves to the top of a nearby prominence, such as a stonewall, where it can flee into crevices at any sign of danger. Being nervous little animals, there often will be a commotion of front and hind prints at any feeding spot soft enough to record tracks.

Track and Trail Comparisons

Red squirrel: There is little in the appearance of a chipmunk's tracks to distinguish them from those of a red squirrel except their slightly smaller size. Even then, measurement of very small features in the field often involves fairly large error and so a margin of only an eighth of an inch difference in hind print width can be difficult to discern.

Differences in the chipmunk's track patterns are more helpful. In the trapezoidal four-print bounding pattern common to all members of this family, the hind straddle of a chipmunk is normally about 2½ inches, about an inch narrower than the usual pattern width for a red squirrel. However, both species have flexible hind legs and can spread or close them beyond the expected parameters. And when a chipmunk's tracks and trail in snow are enlarged by melting, in a spring thaw for instance, other evidence must be looked for. Chipmunks tend to hold their front feet parallel or splayed while red squirrels often give the appearance of toeing them in. Red squirrels are more of an arboreal species, usually spending less time on the ground than chipmunks, whose foraging patterns with very short pattern length are more usual compared to the long patterns of a red squirrel speeding from tree to tree. None of this is very precise, of course, and both species climb trees to harvest ripening nuts. Any ambiguously sized patterns found in the dead of winter in the Northeast can be safely attributed to a red squirrel, since chipmunks are asleep underground at this season.

Flying squirrel: The tracks and patterns of both species of flying squirrels are around chipmunk size and might be difficult to tell from them were it not for the fact that they are not nearly as often found. However common these little squirrels may be in eastern woodlands, they are mostly arboreal in their behavior, gliding from tree to tree without touching the ground except in openings too wide to glide all the way across. At night they may come to the ground to gather food, and their tracks may also be found in snow near bird feeders where they take full advantage of human generosity. Since, unlike chipmunks, they do not hibernate, any chipmunk-size squirrel trails in snow in the dead of winter may safely be attributed to them.

Scat

There is little to differentiate chipmunk scat from that of other squirrels except size. Figure 23.1 D shows a small collection of granular pellets of typical size and shape, collected in Chelmsford, Massachusetts, and Conway, New Hampshire. In both instances the color was black. As chipmunks spend a lot of time underground, much of their scat will be hidden from view. However, searching through a collection of shells at a feeding site on a stonewall or stump may reveal one or two.

Sign

Larger squirrels must totally dismantle an acorn shell to get at the meat whereas a chipmunk with its smaller mouth may need only to open about half, as in Figure 23.1 E, a collection photographed at a typical site in Brownfield, Maine, on top of a stone wall at the edge of a pasture. A scat of appropriate size found on a stone wall next to acorns with openings such as these and with the alarm "chucks" and chatters of the resident sounding nearby can safely be attributed to this animal. Although chipmunks may feed more than once at the same site on a stone wall, the accumulation of debris rarely amounts to the size of a red squirrel's midden.

Chipmunks are great diggers, constructing an elaborate system of tunnels for safety, birthing, storage and hibernation. The spoil from the digging is ejected from a construction hole and forms a large and conspicuous fan of subsoil that is easily spotted on a slope of contrasting color like oak leaf or pine needle ground cover. Once construction or renovation is completed, the chipmunk plugs this hole and uses more discreet entrances that it has dug from the inside out in order to leave no debris. These may be hidden under rock walls, but apparently the animal has a faulty idea of where the tunnel will come to the surface as occasionally the hole will be in the middle of a hiking trail or other inappropriate location. The tunnel entrances can be distinguished from those of woodland voles by a larger diameter of about 2 inches and by a tendency for the tunnel to drop straight down for a foot or less before taking a right angle turn to the horizontal. This is probably a design feature intended as a defense against weasels, which may have trouble making the turn with their long bodies. Probing the tunnel entrance with a pliable stick will reveal this angulation for you. Woodland vole tunnels, on the other hand, have narrower entrances of perhaps 1 to 1½ inches and are usually near the surface, although I once found one nearly 8 inches deep in sandy soil. Mole tunnels are usually found in fields or other earthworm-rich habitats with renovation spoil in the familiar mound.

Habitat

Eastern chipmunks inhabit broad-leafed woodlands and woodland edges in most of the eastern states and some in the Midwest. In the boreal forests of the north, the red squirrel replaces both this species and the gray squirrel. Thus any ambiguously sized print or patterns found in deep conifers was made by a red squirrel or, more rarely, a flying squirrel. Chipmunks will consume almost anything edible that they can find and handle in their range, both animate and inanimate. In New England they are most often seen near the many stone walls that run through woodlands reclaimed from pasture after the decline of agriculture in the mid-19th Century. They also make good use of food scraps that they find in campgrounds and picnic areas.

Although they are omnivorous, I have long suspected that chipmunks have mouthparts specially designed for opening hard ovals, since prominent among their broad range of foods are nuts, beetle backs, small turtle shells and birds' eggs. I once came upon a chipmunk besieging a chickadee nest in a dead stub which was being defended solely by the female. Nearby I found a dead bird that I believe was the male, apparently trapped in the nest, whose brainpan had been neatly removed while the rest of the carcass was ignored. (Against all odds the female successfully defended the eggs and fledged the brood.)

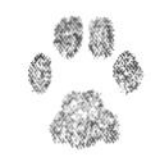

Trackard 24 – Mouse

In the United States there are over 50 species of mice, most of which have very limited ranges. Among the most widely distributed are the white-footed and deer mice. These are two native species that may find their way from nearby woodlands into your attic and cupboard in winter and so may be more familiar to you than you might wish. Since their ranges, habits and habitats overlap, they will be dealt with here as a single species. Because the tracks and trails of most species of mice are similar, the information presented may be extrapolated to other mice, as well. Knowledge of which mice live in your area as well as any peculiar habitat requirements may help in efforts to identify them.

The tracks and trails of the two species of jumping mice are quite different from those of other mice and will be dealt with separately at the end of the chapter.

Mice are nocturnal creatures with big eyes and ears to warn them of danger in the dark. They nest and tunnel in enclosed spaces, becoming active after dark when they move abroad in search of seeds or other edible matter. These they may consume on the spot or store, sometimes in large quantities, in a cache underground or in a hollow.

Tracks

Since both of these closely-related species forage overland at night, their tracks are often found in natural areas in both mud and snow. Having similar feet and anatomies they leave behind tracks and trails that are hard to differentiate. A track width of only about a centimeter means that the pads of white-footed/deer mice are rarely discernible except in very good surfaces such as smooth mud or thin wet snow over dark ground. This latter was the case in Figure 24.1 A, photographed at Concord, Massachusetts. Like most rodents, mice have five toes on the hind foot and four registering toes on the front. The overall effect of a mouse's four-print bounding pattern is that of

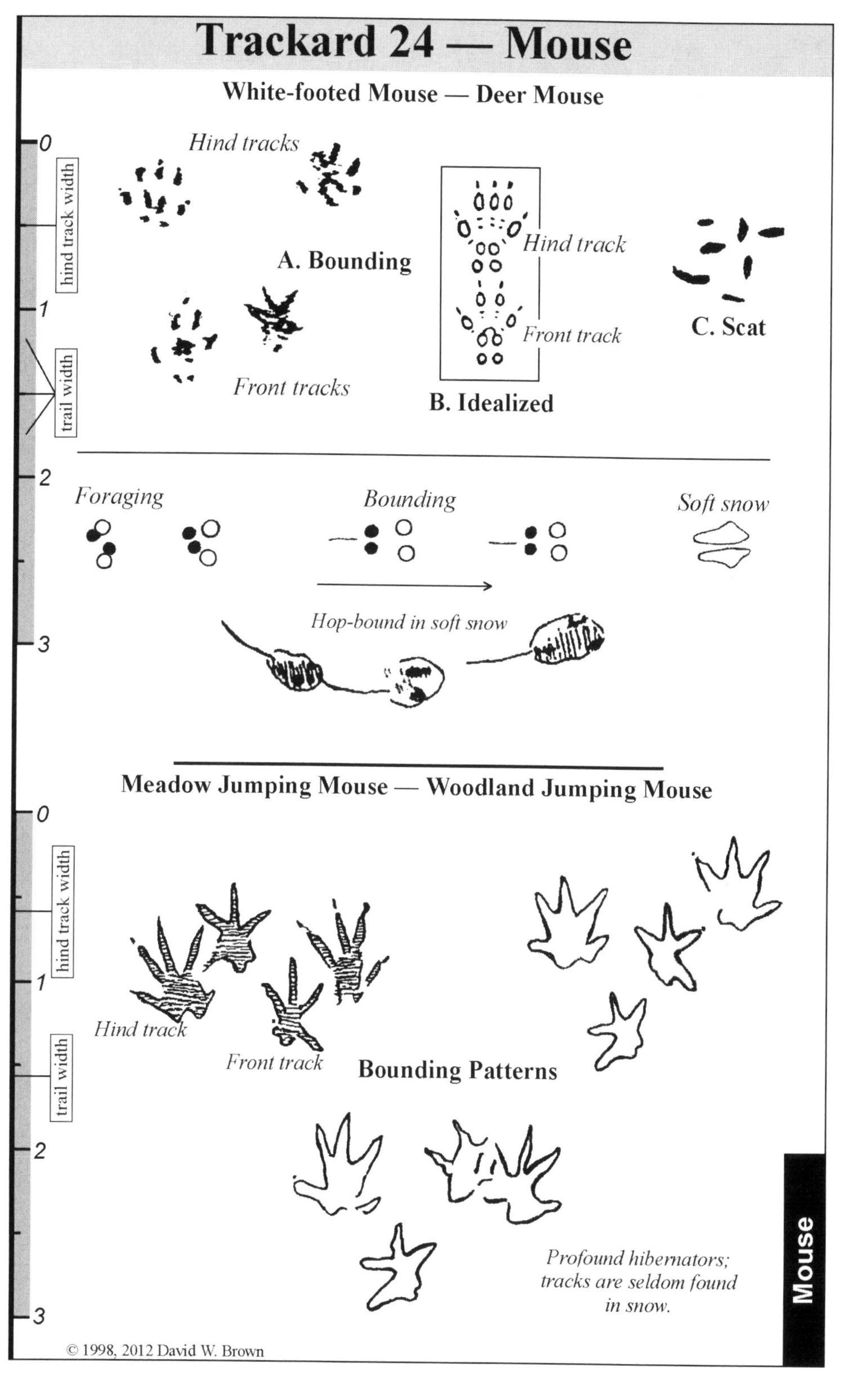

Figure 24.1. Trackard 24 – Mouse.

a tiny squirrel. Other small mammals with which mouse tracks might be confused are shrews and moles, but members of those families have five toes on both front and hind feet.

The hind print shows five toes, with the pads of the three central toes more or less evenly arranged. The medial and lateral toes are commonly splayed and are shorter than the central toes. Secondary pads register as four dots while the staggered tertiary pads are farther up the foot and seldom show in a print.

The front print shows four toes. Three secondary pads form a triangle of dots closely grouped. In very clear prints they can have the characteristic appearance of two balls hanging from an arc, as shown in Figure 24.1 B, idealized from a photograph of tracks in mud taken at Brownfield, Maine. Tertiary pads often register in front prints, appearing as two dots to the rear of the secondary group.

Trails

When traveling any distance, mice use a bounding gait. Adapted to moving over the surface without the overhead interference of a tunnel ceiling, they can afford this gait with its high arc of stride in contrast to the flat scurrying gait of voles that live in tunnels and covered runways. The resulting pattern for these mice, when they move over a firm surface at least, is a trapezoid of prints as shown in Figure 24.1 A and in bounding patterns in the middle of the upper line of the trails section. The pattern width is generally between 1¼ and 1½ inches. Hind feet are advanced and even; front prints are even, retarded and have a narrower straddle. When a mouse slows down to forage, the hind prints don't overstep so much, and so the front prints will register only a little behind, between or even a little ahead of the hind in each pattern, as shown on the left end of that track line on the card.

Tail marks may or may not show depending on the species and the surface. In some areas of the country deer mice are reported to have longer tails than white-footed mice, but whether or not a tail mark shows in a pattern depends more on the softness of the surface, the arc of stride and the speed of the animal than it does on its tail length.

In firm snow these mice often go on seemingly suicidal nocturnal perambulations, leaving long trails of trapezoidal 4X patterns. This behavior may sometimes be explained by the fact that pendent winter seed structures of birches, as they disintegrate slowly over the winter, scatter their minute seeds on the snow. In crusted conditions the wind blows them over a large area inviting risky wandering by mice. On the other hand, sometimes the direction of the trail seems more arbitrary. It may begin at a small hole at the base of a shrub, for instance, travel in a more or less straight line for as much as a quarter of a mile, without showing any foraging behavior, and then disappear into another hole at the end. The frequency of mouse bones found in owl pellets testifies to the hazards of this sort of behavior.

In soft snow where the mouse sinks in more than an inch, it will revert to a hop-bounding gait where it lands on its front feet and then brings its hind feet down on top or just slightly ahead of them in a direct or indirect registration. A pause occurs while the mouse extricates front feet from under hind, after which it uses the packed snow, firmed by both sets of feet, to push itself off with its strong hind legs into the next bounding arc. This is the hop-bounding pattern also used by squirrels in soft, deep conditions. An example of the resulting 2-track patterns is shown on the bottom line in the trails section of Trackard 24, drawn from a photograph taken at Lincoln, Massachusetts. In slightly shallower snow a signature profile of opposing triangles may be found, which is illustrated at the upper right of the trails section. This profile of opposed triangles is shared with squirrels in similar conditions, differing from them only in size.

Track and Trail Comparisons

Voles: Voles are mouse-size creatures with small ears and eyes that spend most of the winter under the snow. The structure of their feet is similar to mice. However, clear tracks are seldom found because these animals spend most of their time under grass, leaf litter or snow. When they appear on the surface, track detail is often obscured by comings and goings over a narrow route. The length of their trails is usually short and often under some sort of overhead cover such as a fallen branch. One or more animals may pass over the trail repeatedly, packing it down into a trough. Mice rarely follow such set routes; usually their trails are of one animal traveling by itself or foraging out in the open. Even when a mouse travels repeatedly from nest to cache and back, it usually takes a slightly different line each time, as if the efficiency of packing down a trough in the snow never occurs to it. When a single vole pass is found, it often contains a variety of patterns in even a short distance: usually scurrying prints arranged on diagonals interspersed with ragged bounding patterns. These contrast sharply with the neatness and regularity of mouse patterns.

Various vole species have different length tails ranging from short to shorter, but the appearance of a tail mark in any small rodent trail is more a function of the depth of the track pattern than the anatomy of the animal. When tail marks do appear in a vole trail, they are apt to be shorter and more intermittent than the more regular and longer marks that mice leave in soft conditions.

Shrews: Shrews have five toes on both front and hind feet as opposed to the 4 front and 5 hind of mice. However, the feet of most shrew species are so small, sometimes smaller than the snowflakes in which they are impressed, that clear prints in which the toes can be counted are seldom found. Although there are a number of species of shrews, varying considerably in size, most make trails with very narrow widths, usually around an inch, compared to the 1¼+-inch width of mice tracks. When a shrew's 4X bound pattern is found, it usually appears as a very shallow trapezoid as opposed to the more squared-off patterns of traveling white-footed/deer mice. Like voles but unlike mice, shrews often pass back and forth over the same short route, wearing down a trough in the snow. The appearance of their trails when they do wander on the surface is presented on Trackard 25 (Figure 25.1).

Scat

Typical scat is shown in Figure 24.1 C, from Saugus, Massachusetts. The tiny black to dark brown pellets are about the size of bits of pencil lead and are commonly found mixed with the debris that accumulates inside the base of a hollow tree where the mice have been denning.

Sign

No wild animal will do more work than it has to in order to acquire food. When mice open nuts and acorns, they make a hole or holes just large enough to extract the meat. In acorns, this means that the cap is removed and the shell gnawed with very fine incisor marks no more than half way down the acorn. Chipmunks, with larger heads and mouthparts, need to gnaw more of the shell to extract the meat while squirrels generally have to disassemble the shell completely, peeling it into strips. When mice open hickory nuts, which have very hard shells, they usually gnaw one or two very small openings into the meat chambers on either side of the nut. Larger rodents must gnaw large ragged openings.

Mice, like other herbivores, have two long central incisors separated by a wide gap from their molars. These incisors make marks that can often be detected on the bones of animals that have died over the winter. Deer bones and shed antlers often show the tiny parallel marks of mice seeking calcium for their diet with which to maintain their own bones and build those of gestating offspring.

Dens

Mice are solitary animals that, nevertheless, often den together in huddled groups for mutual warmth in the winter. Such dens frequently are inside hollow trees packed with shredded leaves and grass for insulation. Another favorite denning site is an abandoned bird nest. A cap of fine material is added and a hole chewed into one side, resulting in a snug recycled home. When snowfall adds a white cap of its own to the nest, the insulating capacity of soft snow makes it even cozier within, at least until the snow melts.

Jumping Mice

Woodland jumping mice and meadow jumping mice are two look-alike species that have tails so long that they remind us of the more celebrated kangaroo rat. Since they are profound hibernators, their trails are seldom seen in snow but may be encountered in damp bottomland soil or mud puddles in warmer seasons. The illustrations on Trackard 24 (Figure 24.1) were traced from plaster casts of several patterns found on a muddy dirt road in Brownfield, Maine. Note the large size and long toes of the hind prints with the medial and lateral toes attached to the foot well back from the base of the central three. The reliability of the medial fifth toe registering in the print is marginal. Sometimes it appears vestigial, or withdrawn slightly from the print, or it is obscured by the front prints.

Note also the extreme splay of the front feet whose central axes point away radically from the direction of travel. These often look similar to me to the front prints of a frog except that they are oriented outward rather than inward. Despite the larger size of jumping mice feet, the pattern width (hind straddle) is still a mouse-like 1½ inches.

The grouping of tracks in the patterns displayed is significant. Among other rodents such a tight grouping of front and hind prints would suggest foraging behavior, but with jumping mice they appear in isolation, with long inter-pattern distances. This suggests that the animal was bounding, or "jumping" through a series of high arcs, not foraging in front of its nose for food.

Trackard 25 – Vole and Shrew

Voles and shrews are combined on Trackard 25 (Figure 25.1) partly because members of both of these families are small mammals that make similar tunnels or runways. At least one species of shrew, the "short-tailed," moreover, often invades the runways of meadow voles to prey upon them, so that the sign of the two may coincide. Each group is treated separately here.

Vole

Voles should not be confused with "moles," the dark gray lawn-tunnelers with huge front feet (see Trackard 26 – Miscellaneous). Voles are a group of rodent species similar to mice but with small eyes and ears, adapted to moving in the dark through narrow runways under snow, earth, leaf litter or overhanging grass. About thirty species occur in North America, with a half-dozen occurring in the Northeast, including meadow voles, red-backed voles, woodland (or "pine") voles and southern bog lemmings. Yellow-nosed (or "rock") voles and northern bog lemmings occur in northern New England. Each of these species prefers a slightly different habitat although there is considerable overlap. Voles tend to be active around the clock, and some species breed through most of the year. This latter is significant because voles are staple prey for hawks, owls, foxes, weasels and other predators.

Tracks

The footprints of voles are similar to mice in size and other details: four toes on the front foot and five on the hind, arranged more or less symmetrically and with secondary pads also balanced laterally. A representative collection of front and hind prints are presented in Figure 25.1 A, drawn from a cast made at Brownfield, Maine. The size of these prints will differ slightly, corresponding to the varying sizes of the different species.

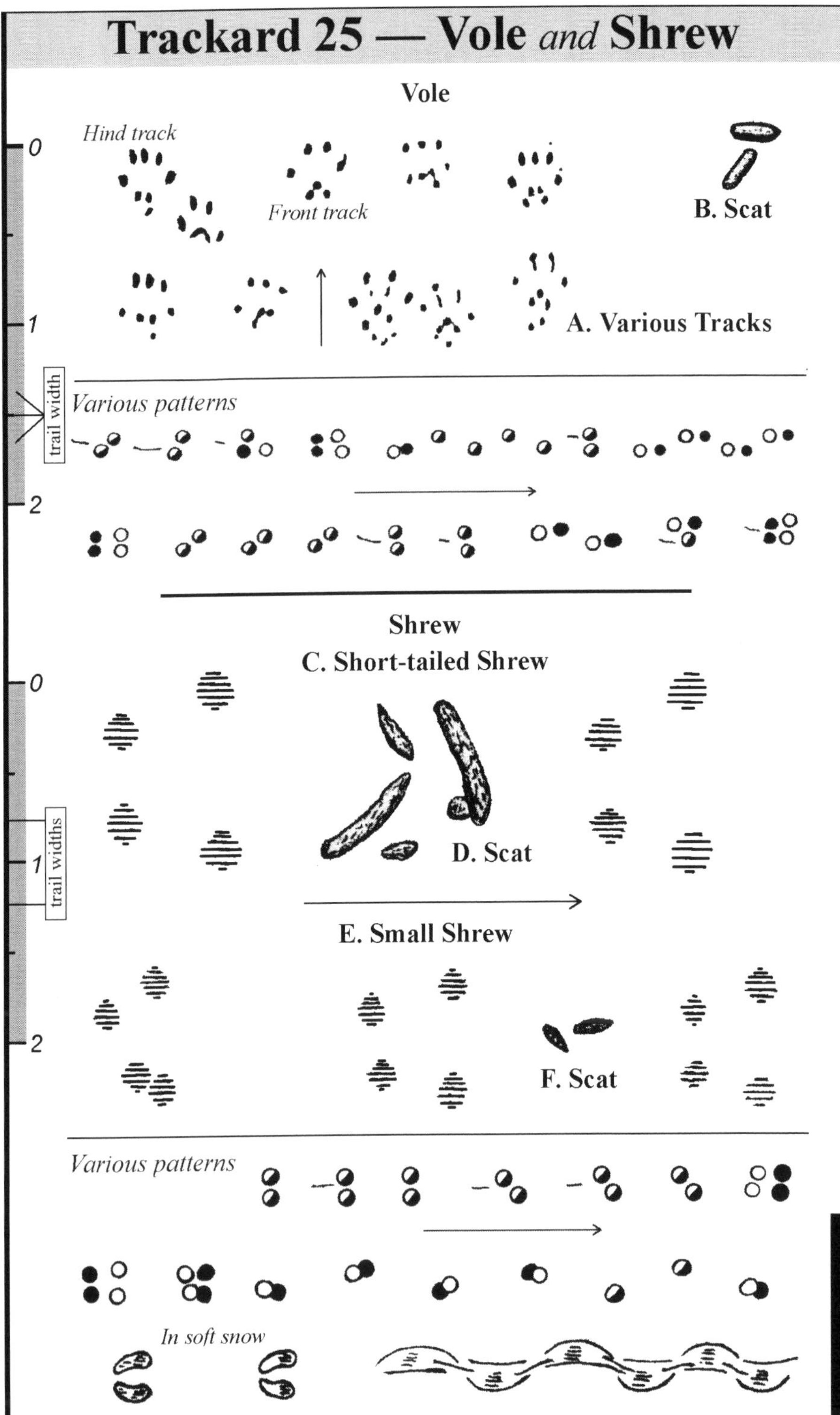

Figure 25.1. Trackard 25 – Vole and Shrew.

Trails

Although the various species of voles differ slightly in size, distinguishing among their trails on the basis of measurement of things so small is impossible. The trails most often found are those of meadow voles, which occasionally come to the surface in areas with thick grass such as abandoned fields and meadows.

Vole identification is almost always made by track pattern rather than by the appearance of individual prints. Where mice tend to leave neat 4X and 2X bounding patterns in snow, vole trails are more irregular. Mice spend a good deal of time on the surface where bounding is an efficient means of getting around, but the tighter confines of runways and tunnels in which voles spend most of their time instill the habit of a flatter scurrying gait. On occasions when they do travel over the surface of snow, they continue the scurrying habit interspersed with occasional bounding. Their lack of practice with bounding tends to show in ragged, uneven 2X, 3X and 4X patterns following one another without consistency. The shallow patterns shown in the trails section of Trackard 25 are a composite from Lincoln, Massachusetts, and Glen, New Hampshire, recorded on fairly firm show. Figure 25.2 I shows another trail from Lincoln. In this case, as the winter was drawing to a close, rain had percolated down through the snowpack and frozen at night next to the ground. In the winter, voles travel in tunnels and spaces at the bottom of the snowpack. As these were filled with ice at the time, this vole was forced to the surface where a dusting of fresh snow recorded its trail along with that of one of its principle predators, a red fox, whose track appears near the left end of the trail.

On other occasions some undersnow obstacle such as a fallen log or rock may also force voles briefly to the surface. Nonetheless, they are shy of skyspace, as might well be appreciated, so the resulting trough of tracks may follow overhanging branches or stems bent to the ground by winter storms. Shortly, such trails will enter a hole in the snow.

Although voles are generally pictured as rotund, this shape is postural, intended for reducing heat loss. When entering a hole, they are capable of lengthening their bodies and compressing their girth so as to make an amazingly small entrance. One such, made by a meadow vole in settled snow that I recorded in Glen was well under an inch in diameter, although the usual is between $1\frac{1}{8}$ and $1\frac{1}{4}$ inches. It is in the interest of voles to make as small an entrance to their

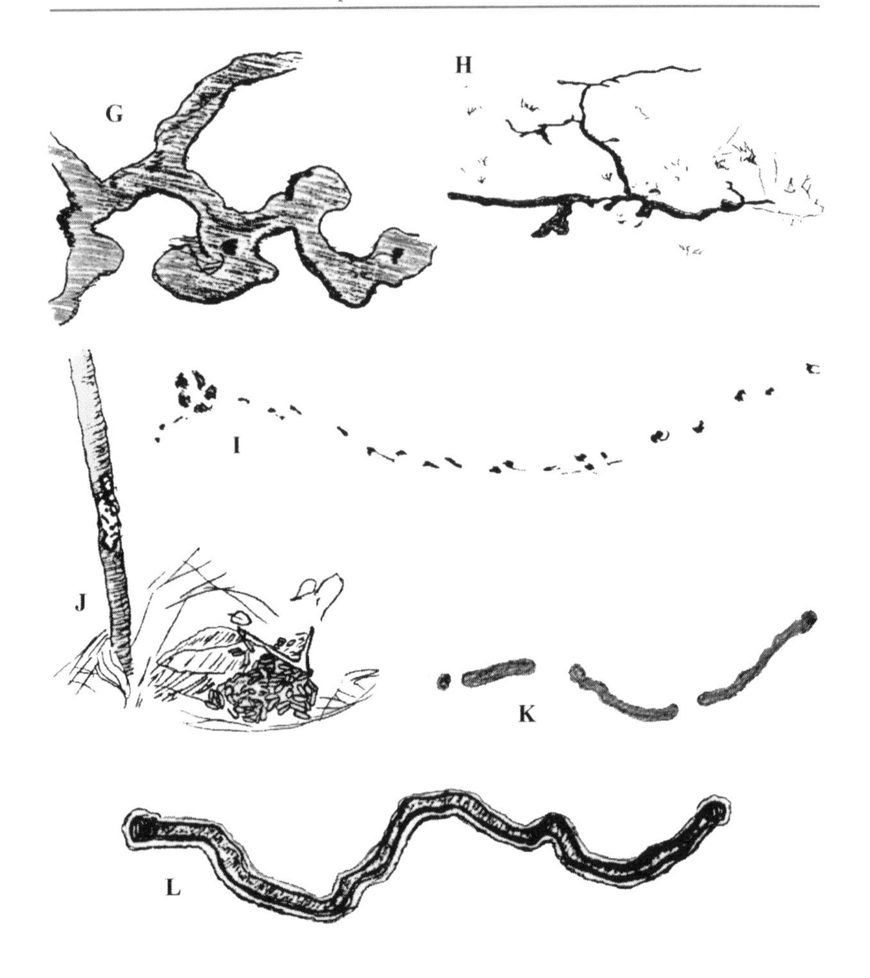

Figure 25.2. Vole and shrew sign, all greatly reduced.

G. Meadow vole mud runaways, cores and pits exposed in a field by spring thaw. Carlisle, Massachusetts.

H. Vole runaways in short grass under snow, exposed by the thaw. Princeton, Massachusetts.

I. Irregular scurrying trail of meadow vole on snow in a field. Note the fox print at left. Lincoln, Massachusetts.

J. Bark chewing and large pile of scat where a meadow vole had fed under the snow for some time. Lincoln, Massachusetts.

K. Shrew tunnel in snow exposed intermittently in late winter thaw.

L. Trough in snow made by shrew crossing from hole to hole on a bridle path. The trough was 1-inch wide and contained both scat and a tiny urine mark. Concord, Massachusetts.

undersnow passageways as possible since weasels have an anatomy designed for entering tunnels in pursuit of prey.

Different vole species have tails of varying lengths that may or may not appear in the trail. In a soft surface or in a bounding gait, the longer-tailed species such as meadow voles or boreal red-backs often show these marks as illustrated in the trails at the bottom of the vole section on Trackard 25. Note that these marks are much shorter than one typically finds in a mouse's trail.

Scat

Voles are coprophagous, that is, they reingest their scats, running them through their digestive system a second time to extract all possible nutrition. The result is scat of very fine texture and uniform size. The perfect little "sausages" illustrated in Figure 25.1 B, found in a grass runway at Princeton, Massachusetts, look quite different from the more irregular droppings of mice. The group in Figure 25.2 J was found at the base of a gnawed red-osier dogwood where a meadow vole had sat under the snow for some time, chewing bark and depositing in the process a large clump of scat. The spring thaw revealed the scene with reddish scat soaked pink by melt water. When feeding on other kinds of bark and rootlets, fresh meadow vole droppings appear brown, or greenish brown if fresh grass is in their diet as well.

Sign

In summer, meadow vole runways may be found by lying on the ground in a field and searching at eye-level for bowers of grass bent over to conceal the passing animal from predators above. After the grass has been beaten down by storms, look for a hole in the mat of grass wide enough to insert two fingers. Sometimes this hole may be disguised by lengths of grass, cut off evenly and arranged over it. Pull the grass apart and underneath you will find a maze of passageways worn into the dirt by continuous passage of tiny feet. Along these runways you will find cut grass blades here and there and perhaps scat as well.

Once spring thaw is complete, the winter runways of voles are often seen as irregular muddy troughs wandering over the surface with occasional holes down into the ground. The network of these runways can be quite extensive. A short section of one found in March at Carlisle, Massachusetts, is shown in Figure 25.2 G.

Woodland voles, which typically tunnel in leaf litter, are usually detected where their trails cross danger areas, open places such as human foot trails that hawks and owls use as hunting lanes. Here this species of vole, which is better adapted to digging than others, will construct a shallow tunnel the hump of which may be seen on the surface. This tunnel will be about two fingers wide and will disappear into the leaf litter on either side of the foot trail.

While exploring a field in the winter one may find occasional holes in the snow about an inch across but without tracks leading into or out of them. Enlarging one with a finger will show that it leads to a vole runway at the base of the snowpack. There has been some debate about why voles make these openings, especially since they tip off the presence of the runway to a predator moving over the surface. The traditional theory is that voles make them as ventilation shafts in order to exhaust carbon dioxide that accumulates under the snow in the winter as a byproduct of both animal and plant respiration.

Sign Comparisons

Shrew: When shrews make passageways in grass, they press back the grass rather than biting it off. Both shrews and moles are insectivores, lacking the two long central incisors needed to nip grass or to gnaw bark, for that matter.

During spring thaws, irregular tunneling in the snow is often discovered that could be attributed to either voles or shrews. Voles tend to spend the winter at the interface of ground and snow, however, while shrews often tunnel just under the snow surface. A tunnel maze exposed by a thaw, then, which shows bare ground, such as the one in Figure 25.2 H, is more likely to be a vole tunnel than a shrew tunnel. To confirm this, search carefully along it for cutting on grass blades or other green material that has been insulated under the snow during the winter.

At least one large species of shrew, the short-tailed shrew, preys on meadow voles, often pursuing them through their passageways. Once a vole has been dispatched, the shrew may take over the tunnel system for its own purposes. Distinguishing the two in these circumstances is nearly impossible unless scat is found. Compare Figure 25.1 B and D for expected differences in appearance of the scat of these two animals.

Dens

Meadow vole nests in winter are spheres of grass about the size of a softball, formed by the animal under the snow. The insulated interior is used for resting as well as birthing. By contrast, woodland vole dens are usually in the rotted root structure of an old stump where the earth is so spongy that it gives easily under a person's foot. These locations are often riddled with tunnels and openings. Patiently sitting downwind near one may bring the animal into view since voles are intermittently active both day and night. Such places are also used by red squirrels for caching cones, but when a squirrel is the digger, the holes will be much wider, about 3 inches to about 1½ inches for the vole, and a pile of stripped cone scales will be found somewhere in the area.

Habitat

Although there is much overlap in the habitats of various vole species, meadow voles typically live in fields with tall grass or in wetlands with the same. Woodland voles tunnel through the leaf litter of upland forests. Boreal red-backs are most often found in low slope situations, and the bog lemming's favorite habitat is in its name. Despite a high birth rate and abundant sign of their presence, voles are unfamiliar to most people because they stay out of sight under grass, snow, leaf-fall or earth. Any trail of a small animal wandering over the surface for any distance at all will either have been made by a foraging mouse or by a shrew that emerged onto the surface of the snow and then, with its poor vision, couldn't find its way back under. On the other hand, any vole trail will disappear back under something as soon as possible.

Shrew

The shrew family comprises a large number of species that share certain characteristics. Most of them are smaller than mice, voles or moles are generally more common than any of these species in woodland habitats. Like moles, to which they are closely related, they have poor eyesight and very small ears, features of their subsurface existence where the first is unneeded and the second is a hindrance in the tight confines of a tunnel. Their most distinguishing feature is a long pointed snout that results from their dentition. Shrews are carnivores, or more usually insectivores, that feed on small, living animals. Like others of this sort, they have prominent central incisors followed by a row of unicuspid pre-molars with which they seize and hang on to a struggling victim. Herbivores like voles and mice, on the other hand, have long and sharp central incisors, then a gap with no canine teeth or pre-molars and then a set of grinding molars, all characteristics adapted to cutting and chewing vegetation.

Tracks

Shrews have five toes on both front and hind feet, contrasting with the rodent arrangement of 4 and 5. The resulting tracks, however, are usually so tiny that detail is hard to see, especially since most shrew tracks are found in snow where the foot imprint may be smaller than the snow crystals in which it is imbedded.

Trails

Shrews range in size from the little masked and pygmy shrews that are about the size of a man's thumb and weigh a fraction of an ounce up to the larger shrews such as the short-tailed, which is nearly mouse size. This range is represented on Trackard 25 by the bounding patterns, actual size, of the little masked shrew in Figure 25.1 E and the larger short-tailed shrew in Figure 25.1 C. Whatever the size of the individual species, the width of these patterns is distributed around 1 inch, varying only about ¼ inch either way. Conveniently, my thumb is exactly an inch wide and provides a quick field reference for the width of patterns, tunnels and troughs created by these little predators. Mouse and vole patterns, by contrast, are generally about the width of two of my fingers, another quick field reference.

Identification of shrews usually depends on the characteristics of pattern and trail left behind in snow by the passage of the animal. In soft conditions where the depth of the tracks is about a centimeter or more, shrews respond in the same way as many rodents and mustelids, that is, they hop-bound so that hind feet land in the snow pre-packed by the front, leaving a series of 2X direct registrations such as those shown on the top line in the trails section of Trackard 25.

In firmer conditions shrews may bound, leaving the 4X patterns shown in Figure 25.1 C and E. In the case of heavier shrews like the short-tailed, trails of these patterns may be obvious on the snow, but with the tiny masked or pigmy shrews the patterns may be nearly invisible even on the softest surface.

I have long suspected that, unlike rabbits, squirrels and mice, all of which leave a similar bounding pattern, the front prints of a shrew register ahead of and wider apart than the hind. If this is so, it may be a function of the broad, mole-like chest and narrow pelvis of shrews, just the opposite anatomical disposition to that of rodents. With their pudgy little bodies, shrews may just be able to get their hind feet up behind the front, resulting in a series of W-patterns, on average shallower longitudinally than those of other bounding animals. Someday someone in a laboratory with a high-speed camera and a cooperative shrew may prove or disprove this. However it would seem to explain the squat appearance of shrew patterns as well as the relative straddles of front and hind prints. Tail marks may or may not register depending on the softness of the surface relative to the weight of the shrew species involved.

When shrews are running back and forth in packed troughs indented with their own footprints in a jumble of unreadable patterns, they often scurry. This flatter gait is more appropriate for the irregular footing of such a surface. Scurrying, which is actually a quick walk or trot, seems to be their normal gait on bare ground as well, where the tiny animal hurries along, making an occasional hop over irregularities it encounters along the way. If such a trail can be isolated, it will look like the right end of the middle line in the trails section of Trackard 25, the familiar series of direct and indirect registrations, alternating sides that usually denotes a walking gait.

Figure 25.2 L shows a short trail where the shrew repeatedly emerged from one hole and disappeared into another only a couple of feet away. Apparently some sub-nivean obstacle forced it to do this. Notable in this short section of trail is its irregular direction, a clue that the trail-maker was an animal so blind that it was following its nose more than its eyes.

Track and Trail Comparisons

Voles: Most voles tunnel at the snow-ground interface, coming to the surface when their passage is obstructed by ice, a rock or a fallen tree limb at ground level. Many shrew tunnels, by contrast, are just under the surface of the snow and are quickly revealed by a partial thaw. An example of this is Figure 24.2 K. This reflects the different purposes for which the animals create such tunnels. Voles are searching for tender bark and rootlets they find at ground level while shrews often tunnel just under the snow to make a network in which to trap insects, insulated by the snowpack, that drop into it. This doesn't always work since shrews also feed on dormant insects they find at ground level. However, the sort of trail shown in Figure 25.2 K is that of a shrew, while admittedly any tunnel at ground level may be either. Look for grass cuttings, scat and other evidence to tell the difference.

Short-tailed shrews often prey on meadow voles and take over their runways. Since they are nearly as large as their prey, their back and forth trails may be hard to distinguish from those of their victim. However, if they have this carnivorous habit, their scat will have a distinctive appearance as described below. Generally wider than those of most shrews, vole troughs and tunnels measure around 1½ inches wide. Precise measuring of tiny features is difficult in the field and repeated passage along a runway of either animal will eventually widen it slightly. Nevertheless the "rule of thumb" cited above is a useful field reference.

Scat

The scats of shrews may be found occasionally along their runways in the snow along with pale yellow urine stains. The scats presented in Figure 25.1 F were found in an inch-wide runway in Concord, Massachusetts, and represent the smaller end of the distribution. These scats were brown, grainy, and probably composed either of indigestible parts of insects or possibly from vegetable matter the animal was driven by hunger to eat. These scats are smaller than those of voles, but about the same size as mouse scats. Mice have grinding teeth, however, and so their droppings will appear finer textured than scats from shrews, which do not have such teeth. Obviously a hand lens will be helpful for examining things so small in the field.

The larger shrews, like the short-tailed, can leave very different scats than those just described if they are feeding on voles. The deposit shown in Figure 25.1 D was found in a meadow vole runway in a field at Princeton, Massachusetts. The larger pieces were nearly 0.8 inch long and consisted of fur, perhaps that of the host animal or some other vertebrate that uses such runways.

Sign

In addition to snow troughs and tunnels discussed above, shrews also make runways under leaf litter and through grass in the other three seasons. Width, once again, is the main distinguishing feature. It may also help to consider the purpose of the tunnels in the first place. Vole tunnels in dirt, grass or leaf litter are intended either to get the animal across a danger area, say a hiking trail, or to get it to food such as grass stems, tender bark and rootlets. Therefore the tunnel systems that develop express purposefulness, and a search along them will generally reveal a food destination, cut vegetation and appropriate resulting scat. Shrews tunnel more randomly, either in a blind search under the leaf litter for grubs and other insects or to create a network into which these and other small prey creatures will fall. Lacking chisel-like central incisors, shrews press back any vegetation they encounter in the course of their tunneling while voles cut such material. A close look for cuttings along even a short section of runway should tell at least who did the construction if not who is the current resident.

Habitat

Just about any habitat with leaf litter or other ground cover may accommodate a population of these fierce little predators. Shrews by weight have been estimated as the largest part of the mammal biomass in good habitats. However, their commonness may seem to be contradicted by their invisibility. Most are detected when a human alone, eating lunch under a tree perhaps and therefore quiet and immobile, hears a rustling in the leaf litter. A little patience eventually reveals a pointed snout protruding for an instant from among the leaves, followed perhaps by the whole animal as it scurry-hops for a foot or so over the surface before disappearing once again. As invisible as the animals themselves may be, shrew trails over snow are quite common; of tiny animals, only mouse trails will outnumber them. Two examples from a lengthy web of trails in soft snow, recorded near Kancamagus Pass in the White Mountains of New Hampshire, are presented at the extreme bottom of Trackard 25. Shrews are active night and day, but calm, overcast days seem to encourage them to come to the surface and travel out in the open where they can be observed.

Trackard 26 – Miscellaneous

Tracks and sign of many animals, both wild and domestic, other than those covered in the preceding chapters of this book might be found in natural areas and some of these might be confused with similar evidence of their wild counterparts. Included here is a small sampling of tracks and sign of domestic dogs and cats and selected mammals and birds, which might help with the identification of animals not featured in this book. The chapter concludes with a section on the mountain lion, a species with which I have become experienced enough to comment upon only in the last several years. Since there have been increasing reports of mountain lions in areas where they were extirpated many years ago, I thought it would be helpful to present what I know about the tracks, trails and scat of this charismatic creature by way of helping to confirm or disprove reported tracks or sign of its presence.

House Cat

Despite the name, when we let them out at night, cats do not stay at the back door waiting for us to appear in the morning. I often find their tracks deep in the woods and bottomlands as much as a quarter of a mile from the nearest house. Being wilder than domestic dogs, house cats spend the night hunting like their wild relatives. A domestic cat's prints share the characteristics of others of the family Felidae except that they are smaller, measuring usually no more than 1½ inches across with toes closed. The print at the bottom of Trackard 26 (Figure 26.1) as well as Figure 26.2 B shows a representative selection of print profiles in snow and mud. In collapsible snow, a cat may spread its toes for support and measure as wide as 1¾ inches across, in the range of a closed-toed female or otherwise small bobcat. As always, one cannot apply rules arbitrarily but must look at the track and its circumstances. Are the surface conditions such that the toes in a print are spread, enlarging its width? Is this the season when young, undersized bobcats are likely to be abroad?

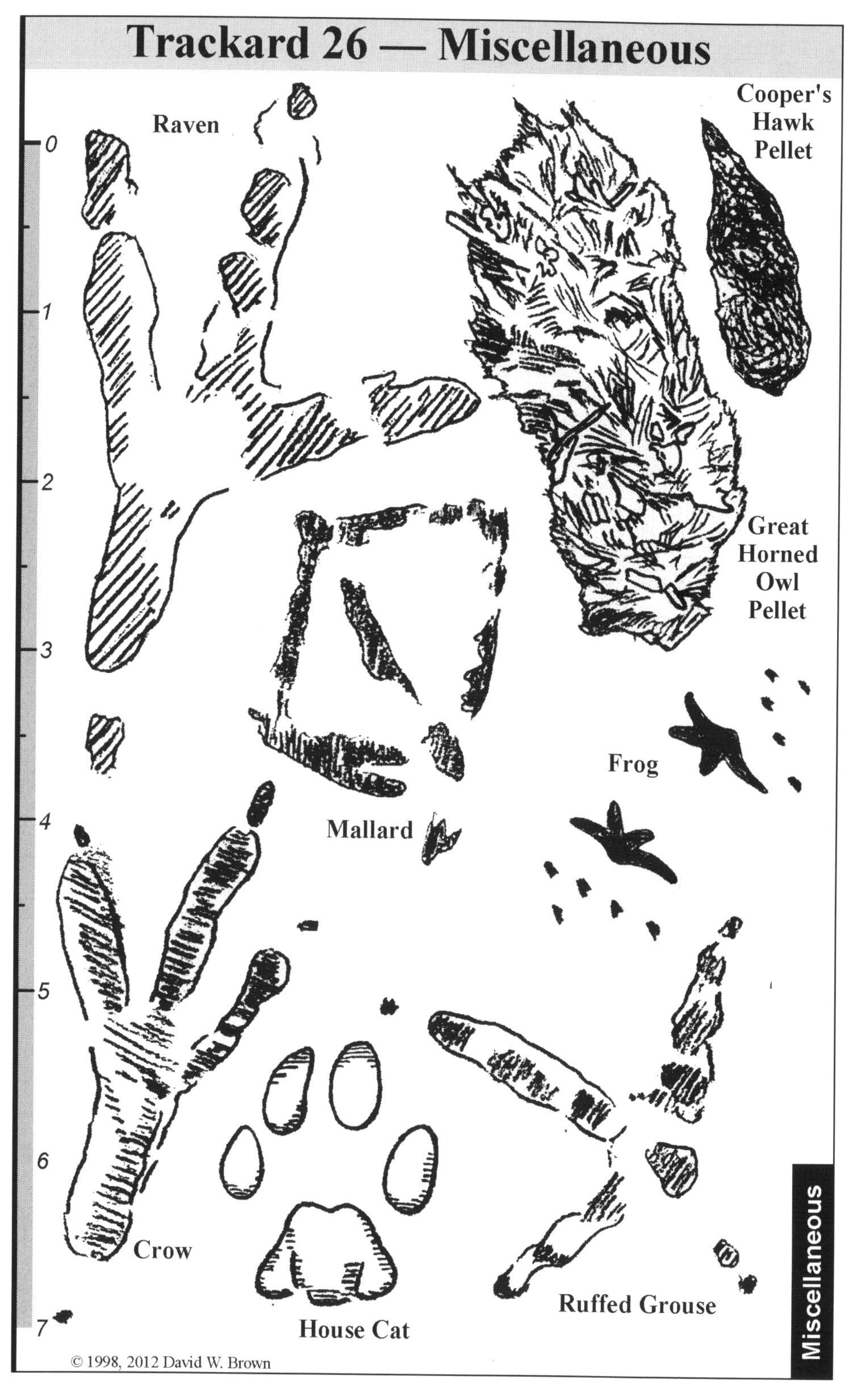

Figure 26.1. Trackard 26 – Miscellaneous.

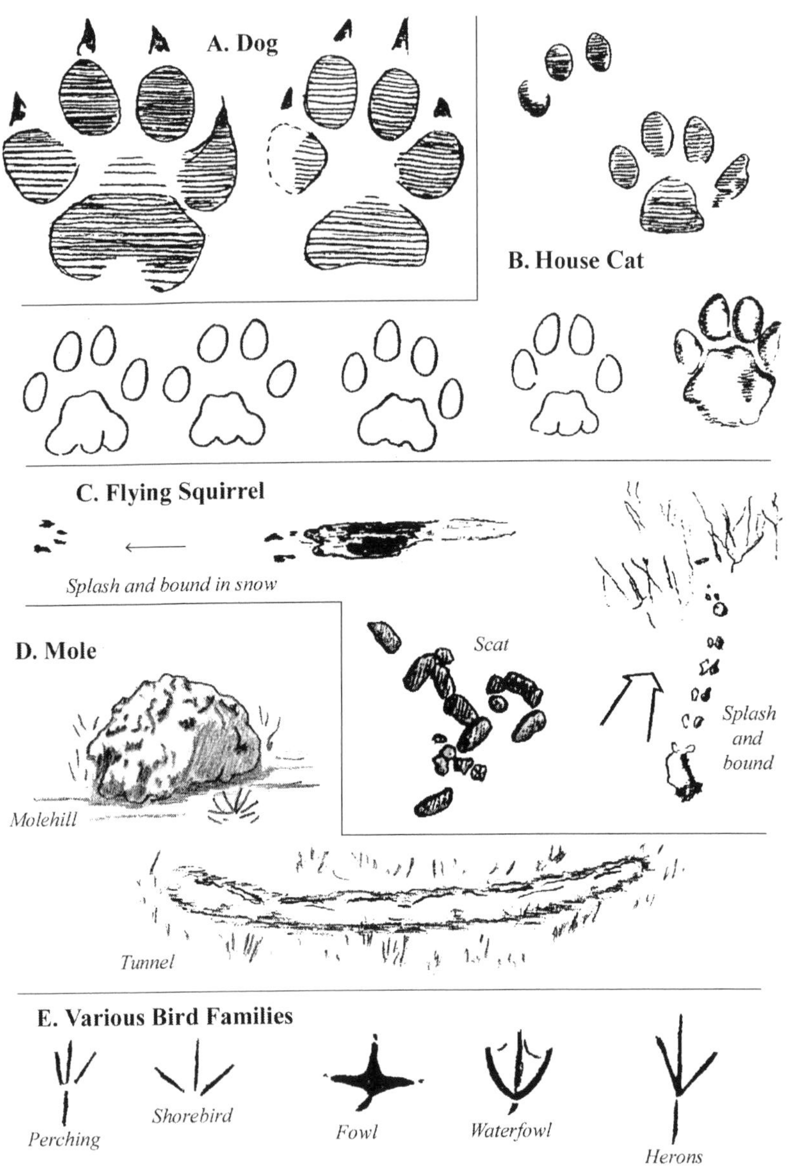

Figure 26.2. Miscellaneous tracks and sign.

A. Domestic dog tracks; front on left, hind on right; ¾ size. Carlisle, Massachusetts.

B. House cat tracks in mud and snow, ¾ size.

C. Flying squirrel trails in snow, greatly reduced; scat ⅔ size.

D. Mole hill and shallow tunnel, greatly reduced. Princeton, Massachusetts.

E. Various bird track profiles, greatly reduced.

Given the wildness of tabby's nature, it is a careful hunter, leaving little information in its trail to distinguish it from its truly wild cousin. Followed long enough, of course, its trail will lead to backyards and residences, which bobcats generally avoid. Even there, however, pet food left on the back stoop or squirrels attracted to seed under a bird feeder may be enough to overcome a hungry bobcat's wariness.

Domestic Dog

Because domestic dogs are descended directly from wolves, their prints show many wolf-like characteristics including, among many breeds, a tendency for the toes to splay, even on firm surfaces where there is no need for extra support. Unlike with red fox, the pads are naked. Nails show prominently unless they have been clipped, a feature that distinguishes Rover from wild canids that dig for a living and so wear down their nails. Figure 26.2 A shows all these characteristics in the front (at left) and hind (at right) prints of the same dog.

Unless they are feral or are running deer, dogs normally stay out of the woods except when they are accompanied by their masters. A shoe print of like age next to a canid print usually means man and dog. Otherwise, domestic dog trails show a wider straddle than either fox or coyote, commonly with indirect and double registrations as well as foot drag. A dog out for a stroll with its owner will charge here and romp there, liberated from the monotony of the house and excited by whatever its senses encounter. It does not need to be efficient about its movements since it will get a can of dog food at the end of the day anyway. The trails of wild canids, on the other hand, especially in the winter, tend to be spare and sober.

When a dog is out with its owner, it will stay with him until something off-trail incites curiosity. It will then angle off to investigate, then angle back to catch up with its master. The resulting right angle trail is an identifying characteristic that may be added to other information at the site. On the other hand, a trail that comes on a line from one direction, crosses the road or trail and continues on in a line into the woods on the other side is more likely that of a wild animal.

Domestic dogs lack not only gait efficiency but also apparently digestive efficiency. Their droppings are familiar to everyone who has walked in a park and found a disgusting mess on his shoe. I have been told that the revolting odor is produced by digestion of grain which is the main constituent of dog food. I also suspect that bacteria proliferating in the relatively rich nutritive medium of domestic dog fecal matter contributes to the odor. Dogs have been bred by humans for many characteristics, but none, as far as I know, for the ability to extract as much nutrition as possible out of its food. By contrast the scat of wild carnivores, whose gut must make the most of what they eat against the constant possibility of starvation, is mild with an almost sweet odor. The odor of domestic dog urine varies with the animal's diet, but does not have the musky skunk-like odor of red fox or coyote. Where domestic dog tracks lead to a musky urine mark, the dog may be counter-marking over the deposit of a wild canid.

Birds

Not all prints in the dirt or snow are made by mammals. This is worth keeping in mind when puzzling over a strange partial track against a confusing background. Most birds have three toes forward and one aft. A representative selection of typical profiles is provided in Figure 26.2 E. Perching bird tracks range in size from little sparrows and juncos that often leave their prints in snow and mud as they forage for fallen seeds, to huge raven prints such as the one that left the print shown on Trackard 26, photographed in snow at Passaconaway, New Hampshire. Also shown is the print of a common crow, a bird closely related to ravens. With their heat-absorbing black plumage, crows are often forced to drop down to the edge of water to drink on summer days, leaving tracks in the mud. The print shown was photographed at the edge of Quabbin Reservoir in New Salem, Massachusetts.

Shorebirds are usually sandpipers, whose prints can be found not only on sand but also in mud along water. Woodcocks and snipe are members of this family, leaving similar prints in soft bottomland earth and wet meadows respectively. Look for the probe marks of a long bill in the vicinity.

The tracks of waterfowl are also found in mud along wetlands and water bodies. The mallard duck print on Trackard 26 was photographed in thin snow over black pond ice at Lincoln, Massachusetts. Note that a duck's footprint viewed looking back along its direction of travel can appear surprisingly like the track of a deer.

Fowl are chicken-like gallinaceous birds such as grouse, pheasant and wild turkey. Their toes are stouter than those of shorebirds, differing from one another mainly in size. The usual distinguishing measurement is along the central toe from the tip (excluding nail) to the base (the point of the arrow formed by the three toes). A ruffed grouse will measure about 2–2½ inches; a pheasant also about 2½ inches but with a proportionately longer middle toe and a wild turkey 3+ inches. In telling the tracks of gallinaceous birds from those of waterfowl where the webbing of the duck or goose does not register, note that the lateral and medial toes of waterfowl tend to curve slightly while those of turkeys, pheasants and grouse are straight.

Birds have a single vent through which they eliminate their droppings, a mixture of fecal matter from their digestive system and urine from their circulatory system. The urine contains urea that marks the dropping with white, usually at one end. This feature can be useful in distinguishing the droppings of a pileated woodpecker feeding on ants, for instance, from a small deposit of skunk scat, which is also commonly composed of indigestible ant parts.

Since their reintroduction into the northeastern states a decade or two ago, wild turkey populations have exploded so that they are now a common sight in fields and along roads. These birds consume a lot of acorns, a major reason for their success in a re-maturing eastern hardwood forest. In searching through the leaf litter for acorns as well as grubs and anything else edible, flocks of these birds tear up large patches of dead leaves, much as do deer, porcupines and bears. A careful look at the scene will show the linear toenail marks of a scratching turkey and perhaps a dropping or two as well.

Carnivorous birds regurgitate pellets of fur, feathers and bones, the indigestible parts of their prey. These may be mistaken for mammal scat but are distinguishable with a little care. Since pellets are coughed up from the bird's gizzard and do not pass through the bacterial atmosphere of the intestines, they lack any dark mucous content or bacterial odor. Owls tend to swallow their prey whole, or at least in as large sections as they can fit down their gullet. As a result their pellets tend to show surface bones and to be more irregular in shape. Hawk pellets, on the other hand, tend to show few bones and have a smoother, rounder, felted appearance. Examples of both are presented on Trackard 26. The owl pellet was found at the base of a large white pine where the bird had roosted during the day, concealed from harassing crows. The hawk pellet was found under a nest around which were perched full-grown juveniles. The presence of the nest and the pellets were given away by splashes of whitewash and natal feathers adorning the bushes. Hawk pellets may also be found near a stump or rock used as a plucking perch to which prey is carried to be eaten. Feathers or mammal tails may be found at the same site. Upon arriving at this perch with fresh prey, the bird may have to cough up a pellet from its last meal before consuming the next.

When an experienced heron catches a fish, it immediately flies to dry ground at the edge of the pond before laying the fish down to spear it. That way the fish cannot revive and dart away into the water. The bird's pellets, deposited there, will contain fish scales and bones, resembling by content and location the scat of river otters. However, each pellet will be a discreet elongated lump separate and of a different age from any others the bird has left there. Otters usually deposit scat in a mass of several different lumps on a small rise in the ground, or they scrape together a mound of vegetation onto which they excrete. As it ages, this scat loses its blackened mucous content, increasing its resemblance to heron deposits. However, heron pellets will be distributed randomly over the site rather than on mounds. A look around may reveal a large splash of whitewash to confirm the identification of bird rather than mammal.

Flying Squirrel

The two species of these handsome, large-eyed animals, the southern flying squirrel and northern flying squirrel, are nocturnal cousins of the diurnal squirrels. Their tracks and trails are not commonly seen since they are more arboreal than the others, sailing from tree to tree without landing on the ground. I have occasionally found their trails on forest roads or ski trails too broad for the little animals to glide all the way across. Characteristically, they land with a splash and then bound the rest of the way to the edge of the clearing where they immediately climb a tree. The two instances illustrated in Figure 26.2 C are from photographs taken on Maple Mountain and Popple Mountain in northern New Hampshire, one at the edge of such a forest road and the other in a small clearcut.

The four-print bounding patterns of flying squirrels resemble those of other sciurids, a trapezoid with the larger hind prints advanced and slightly wider apart than the smaller, retarded front prints. Each such bounding pattern is about chipmunk size, the southern a little smaller and the northern species a little larger, but still not quite as large as that of a red squirrel. Since chipmunks rouse themselves from winter torpor during warm spells, their tracks may be seen occasionally in snow, as may those of red squirrels, which are active all winter. However, neither will show the typical splash-and-bound of a flying squirrel. When flying squirrels forage on the ground, their patterns, so similar in size to other sciurids, may be distinguished by the wider straddle of their front prints. In order to glide between trees, these squirrels need their front feet to be wide apart to support the leading edge of the loose fold that forms their wing. This anatomical feature often causes their front prints to register nearly as wide apart as their hind.

Flying squirrels, especially the short-furred southern species, are group denners in cold weather, huddling together in a tree hole or bird box to pool their warmth. At the base of a den tree one may find a mound of brown scats such as those shown in Figure 26.2 C from a bottomland in Saugus, Massachusetts. In nest cleaning and in other respects, flying squirrels seem to be more fastidious than others of the squirrel family. This includes their work on nuts, especially hickory nuts. While the other squirrels create ragged openings, flying squirrel work is very fine, with a neat beveled opening that looks as if it were made by a precision machine.

Mole

Moles are nearly sightless subterranean mammals that are rarely seen on the surface. As a result, their tracks and trails are not often found. More obvious are tunnels and molehills in fields and lawns. Moles are carnivores, feeding principally on earthworms. As winter approaches, worms tunnel below the frost line to spend the cold months, and the moles follow, renovating their deep winter tunnels into which they hope itinerant worms will fall. The spoil that results from the digging is pushed to the surface, gradually building up into a mound of earth that does not show a hole. Removing the mound will reveal the place where the construction hole is plugged with earth from below. In the spring the worms move back to the surface, as do the pursuing moles, which then set about renovating and digging their warm-weather networks. These are the familiar tunnels that raise a mound of earth above them. Figure 26.2 D shows a mound and a section of tunnel in a grassy meadow. The resulting ridge was 5 inches across and about 2 inches high, with a characteristic cracking of the earth along its spine.

Apparently moles dig their tunnels out to the surface in spots to evacuate waste. On several occasions I have found a pile of soft, black cylinders at the mouth of a tunnel opening that I attribute to this animal. The opening is very discreet, under a scrap of wood, a fallen log or a boulder on the forest floor.

Several species of moles exist in North America, with three being found in the Northeast. Sign of the Eastern mole and the hairy-tailed mole is similar, with the former reported to prefer damper soil. In northern New Hampshire I have occasionally found hairy-tailed moles dead of exposure on the surface of spring snow, a season when their tunnels may flood with meltwater forcing them to the surface where their small body mass is a liability in cold conditions. The third species, the aquatic star-nosed mole, is a bizarre-looking creature that inhabits wetlands as well as dry areas near water.

Mountain Lion

Mountain lions go by many names such as cougar, puma, panther and catamount. Many of these names are regional but nevertheless all of them refer to one species. Figure 26.3 A and B shows hind and front prints respectively of a mountain lion in central Arizona at about half size. The actual measurement for the front print width was $3\frac{3}{8}$ inches and, for the hind, 3 inches even. Both were photographed in shallow sandstone soil with a track depth of less than $\frac{1}{8}$ inch.

Cougars, like other tree-climbing mammals, have deep, rubbery pads that distort easily under load. This leads to a lot of variation in pad shape. The front print in Figure 26.3 B is wider than long, with rounded contours caused by impact as the pads spread under the load of the animal's head and fore-quarters. The leading toe print was somewhat exaggerated by pressure exerted by the animal as it pushed off while walking along a dry wash.

The hind print is longer than wide and more angular, the result of lighter load of the animal's hind quarters. Both tracks show many familiar feline characteristics: lack of nail marks, oval lateral and medial toe-pads rather than the pie-slice appearance of canid prints, a large secondary pad relative to the size of the toe-pads, deep tri-lobing on the posterior of the secondary pad, a flattening on the anterior of the same pad, an aiming off of a longitudinal bisection of the secondary pad in the direction of the shorter of the two central pads that is especially noticeable on the front print and, finally, a pronounced

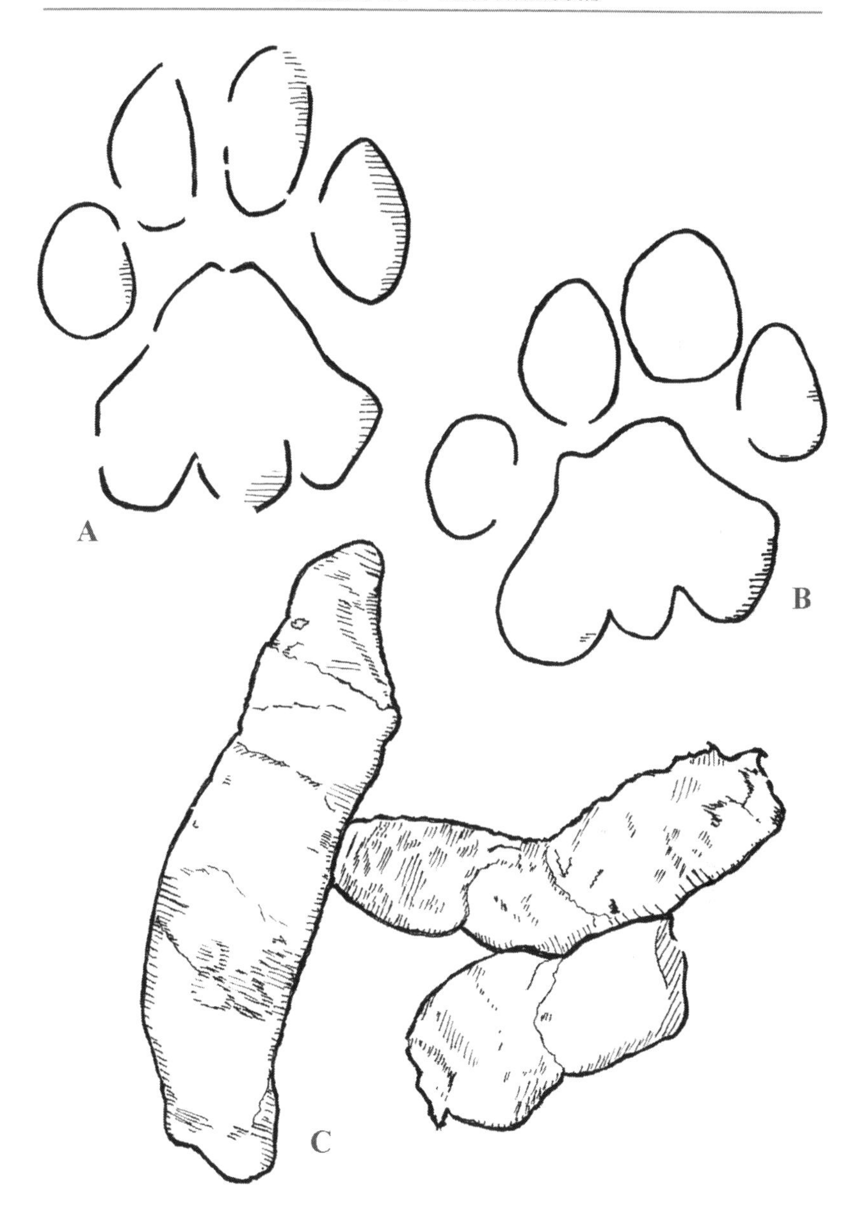

Figure 26.3 – Mountain lion tracks and sign, 5/8 size.

A. Hind print.

B. Left front print.

C. Scat.

asymmetry of the front print with an advanced central toe analogous to the long middle finger of a human.

The deep tri-lobing of the posterior of the secondary pad is especially pronounced. Often in a commotion of other footprints, this is the detail that catches the eye. Very vague prints in loose sand are about the same size and shape of those of range cattle. However the cat shows a short lateral ridge in the center of its track while a cow shows a long longitudinal ridge.

Unless a mountain lion is attacking or escaping, it mostly walks. This may be a direct-register walk in deep snow, but in firmer conditions the cat typically will overstep to some degree. Walking step-lengths that I have measured fall between 14–20 inches. Males are larger than females, however, and the anatomy of tree-climbers facilitates a lot of variation in step-lengths so that the 20-inch upper limit for the cats I have tracked in Arizona is conservative.

Despite the name, mountain lions do not need mountains in their habitat, being as comfortable in the desert as the high country. Mainly they need two things: large prey and a refuge away from man and dog. Their diet is highly carnivorous, depending mainly on elk, deer and peccary. The scat pictured in Figure 26.3 C was found in elk country in the open, grassy pine forests of the Coconino Plateau in north-central Arizona. It was composed of fur and had the familiar sweetish odor of carnivore scat. The maximum diameter was over an inch. Interestingly, it was deposited about a hundred yards from a busy highway, probably at dusk or dawn when the cat was less visible in the low light.

Bibliography

While the information presented in this book on tracks, trails and other sign is original, inevitably the work of others that I have consulted over the years has filtered into the interpretation of that sign. While none of the listed books have been quoted directly, all have contributed to the background knowledge for *The Companion Guide to Trackards for North American Mammals*. I am greatly indebted to them for their experience, scholarship and wisdom.

Burt, William, and Richard Grossenheider. *A Field Guide to the Mammals of America North of Mexico*. Boston, MA: Houghton Mifflin, 1976.

DeGraaf, Richard, and MarikoYamasaki. *New England Wildlife*. Hanover, NH: University Press of New England, 2001.

Elbroch, Mark. *Mammal Tracks & Sign*. Mechanicsburg, PA: Stackpole Books, 2003.

Errington, Paul. *Of Predation and Life*. Ames, IA: Iowa University Press, 1967.

Halfpenny, James. *A Field Guide to Mammal Tracking in North America*. Boulder, CO: Johnson Books, 1986.

Jorgensen, Neil. *A Sierra Club Naturalist's Guide to Southern New England*. San Francisco, CA: Sierra Club Books, 1978.

King, Carolyn. *Natural History of Weasels and Stoats*. Ithaca, NY: Comstock Publishing Associates, 1989.

Kricher, John. *A Field Guide to Ecology of Eastern Forests North America*. Boston, MA: Houghton Mifflin, 1988.

Macdonald, David. *Running with the Fox*. New York, NY: Facts on File Publications, 1987.

Marchand, Peter. *Life in the Cold*. Hanover, NH: University Press of New England, 1987.

Murie, Olaus. *A Field Guide to Animal Tracks*. Boston, MA: Houghton Mifflin, 1954.

Powell, Roger. *The Fisher: Life History, Ecology and Behavior*. Minneapolis, MN: University of Minnesota Press, 1982.

Reid, Fiona. *A Field Guide to Mammals of North America*. Boston, MA: Houghton Mifflin, 2006.

Rezendes, Paul. *Tracking and the Art of Seeing*. Charlotte, VT: Camden House Publishing, 1992.

Whitaker, John, and William Hamilton. *Mammals of the Eastern United States*. Ithaca, NY: Comstock Publishing Associates, 1998.

Williams, Roger. *A Key into the Language of America*. Providence, RI: The Rhode Island and Providence Plantations Tercentenary Committee, 1936.

Index